Managing the Global Nuclear Materials Threat

Managing the Global Nuclear Materials Threat

Policy Recommendations

A Report of the CSIS Project on Global Nuclear Materials Management

Project Chair
Sam Nunn

Project Director
Robert E. Ebel

Task Force Chairs
Graham T. Allison
Roger L. Hagengruber
Roger Howsley
Atsuyuki Suzuki
John J. Taylor

Coordinating Committee Chair
Matthew Bunn

January 2000

About CSIS

The Center for Strategic and International Studies (CSIS), established in 1962, is a private, tax-exempt institution focusing on international public policy issues. Its research is nonpartisan and nonproprietary.

CSIS is dedicated to policy impact. It seeks to inform and shape selected policy decisions in government and the private sector to meet the increasingly complex and difficult global challenges that leaders will confront in the next century. It achieves this mission in four ways: by generating strategic analysis that is anticipatory and interdisciplinary; by convening policymakers and other influential parties to assess key issues; by building structures for policy action; and by developing leaders.

CSIS does not take specific public policy positions. Accordingly, all views, positions, and conclusions expressed in this publication should be understood to be solely those of the authors.

Library of Congress Cataloging-in-Publication Data

Managing the nuclear materials threat : a report of the CSIS Nuclear Materials Management Project / project chair, Sam Nunn ; project director, Robert E. Ebel.
p. cm. — (CSIS panel report)
Includes bibliographical references.
ISBN 0-89206-359-9
1. Nuclear fuels — Management. 2. Nuclear weapons — Materials — Management. 3. Nuclear substances — Management. 4. Nuclear industry — Security measures. 5. Security, International. 6. Nuclear nonproliferation. I. Nunn, Sam. II. Ebel, Robert E. III. CSIS Nuclear Materials Management Project. IV. CSIS panel reports.

TK9400 .M35 1999
327.1'747 — dc21 99-088665

The CSIS Press
Center for Strategic and International Studies
1800 K Street, N.W., Washington, D.C. 20006
Telephone: (202) 887-0200
Fax: (202) 775-3199
E-mail: books@csis.org
Web site: http://www.csis.org/

Contents

Chairman's Statement vii
Participants viii
Executive Summary xiii

Part One

1. Funding Nuclear Security 3
   Appendix A. Current Programs 17
   Appendix B. Sizing the Problem 19
   Appendix C. Obstacles to Nuclear Security Cooperation 21
   Appendix D. Generating New Revenue for Nuclear Security 23
2. An International Spent Fuel Facility and the Russian Nuclear Complex 26
3. Commercializing the Excess Nuclear Defense Infrastructure 38
4. Nuclear Materials Transparency 52
5. U.S. Domestic Infrastructure and the Emerging Nuclear Era 65

Part Two

Introduction 87
6. A View from the Hill 89
7. A View from the Outside 95
8. A View from the Administration 100
9. Looking Ahead 108

Conference Agenda 112
About the Speakers 115

Chairman's Statement

THE END OF THE COLD WAR AND THE DISSOLUTION OF THE SOVIET UNION brought many changes to the world, but none more important and hopefully more lasting than a reduction in the prospect of nuclear war between the two nuclear superpowers. Yet the nuclear standoff that existed before the breakup had also provided a degree of stability, in that a confrontation that could lead to a nuclear clash between the United States and the Soviet Union presented risks clearly unacceptable to both.

Today much of that stability has disappeared, replaced by new challenges of how to avoid the spread of nuclear weapons material and how to keep nuclear weapons out of the hands of terrorist groups and rogue nations. Again we must do all that is possible to reduce these risks.

In our examination of global nuclear materials management that follows, we have been guided by a vision of the world in which all nuclear materials are safe, secure, and accounted for from cradle to grave, with sufficient transparency to assure the world that this is indeed the case. If we are to reduce nuclear arms and bring a halt to nuclear proliferation, effective controls over nuclear warheads and the materials to make them are absolutely essential.

Of the many approaches examined in this report, two stand out in terms of urgency. First, we need a new and comprehensive program embracing additional purchases of highly enriched uranium and plutonium for the purpose of converting these materials to nuclear fuel. Funds earned by Russia could be earmarked to play a major role in helping reduce and stabilize its nuclear complex.

Second in terms of urgency, we must expand our cooperative efforts with Russia to consolidate nuclear materials at fewer locations. These efforts are designed to improve security, accounting, and consolidation and to keep nuclear materials from falling into undesirable hands.

Time is of the essence. We must act, and act now. But the United States cannot do it alone. Broad international cooperation in all aspects of global nuclear materials management is essential. The consequences of failure are far too great and the risks are too high to permit delay.

Sam Nunn
Chairman

Participants

Senior Policy Panel

Graham T. Allison
Douglas Dillon Professor of Government and Director
Belfer Center for Science and International Affairs
John F. Kennedy School of Government, Harvard University

Piet de Klerk
Director, External Relations and Policy Coordination
International Atomic Energy Agency

John M. Deutch
Institute Professor
Massachusetts Institute of Technology

William J. Dircks
Director, Nonproliferation Program
The Atlantic Council

Rolf Ekéus
Ambassador
Embassy of Sweden

Robert L. Gallucci
Dean, School of Foreign Service
Georgetown University

Rose E. Gottemoeller
Assistant Secretary, Nonproliferation and National Security
U.S. Department of Energy

Thomas Graham Jr.
President
Lawyers Alliance for World Security

Roger L. Hagengruber
Senior Vice President, National Security and Arms Control Division
Sandia National Laboratories

Yafei He
Minister-Counselor
Embassy of the People's Republic of China

Roger Howsley
Head of Security Safeguards and International Affairs
British Nuclear Fuels Limited plc

Robert McFarlane
Chairman and Chief Executive Officer
Global Energy Investors L.L.C.

Atsuyuki Suzuki
Professor of Nuclear Engineering
Department of Quantum Engineering and Systems Science
University of Tokyo

John J. Taylor
Former Vice President, Nuclear Power Group
Electric Power Research Institute

Conrado Varotto
Executive and Technical Director
Argentine National Commission on Space Activities

Task Force Members

Task Force I: Funding Nuclear Security
Chair: Graham T. Allison

Matthew Bunn
Harvard University

Gen. William F. Burns
U.S. Army (ret.)

L. Monica Chavez
Pacific-Sierra Research Corporation

Thomas Cochran
Natural Resources Defense Council

John M. Deutch
Massachusetts Institute of Technology

Richard Falkenrath
Harvard University

Steve Fetter
University of Maryland at College Park

Robert L. Gallucci
Georgetown University

William E. Harris
Amarillo National Resource Center for Plutonium

Siegfried Hecker
Los Alamos National Laboratory

John Holdren
Harvard University

Susan Koch
U.S. Department of Defense

Kenneth Luongo
Russian-American Nuclear Security Advisory Council

Eileen Malloy
U.S. Department of Energy

Steve Miller
Harvard University

Ken Myers
Office of Senator Richard G. Lugar

William C. Potter
Monterey Institute of International Studies

Clay Sell
Office of Representative Mac Thornberry

Alexander Stoliarov
Bechtel

John M. Taylor
Sandia National Laboratories

Task Force II: An International Spent Fuel Facility and the Russian Nuclear Complex
Chair: Atsuyuki Suzuki

Edward D. Arthur
Los Alamos National Laboratory

Lake H. Barrett
U.S. Department of Energy

Doyle Batt
Idaho National Engineering and Environmental Laboratory

Harold D. Bengelsdorf
Bengelsdorf, McGoldrick and Associates

Margaret Chu
Sandia National Laboratories

Edward M. Davis
NAC International

Karen Hunsicker
Senate Committee on Energy and Natural Resources

Marilyn F. Meigs
British Nuclear Fuels Limited plc

Robert R. Monroe
Bechtel

Daniel Poneman
Hogan & Hartson, L.L.P.

Alexander Stoliarov
Bechtel

Task Force III: Commercializing the Excess Defense Infrastructure
Chair: Roger Howsley

Willis Bixby
Scientech, Inc.

Madelyn Creedon
Senate Committee on Armed Services

Alex Flint
Senate Subcommittee on Energy and Water Development

John D. Immele
Los Alamos National Laboratory

Fred McGoldrick
Bengelsdorf, McGoldrick and Associates

Thomas L. Sanders
Sandia National Laboratories

Les E. Shephard
Sandia National Laboratories

Alexander Stoliarov
Bechtel

Terrence Surles
Lawrence Livermore National Laboratory

Earl Whiteman
U.S. Department of Energy

Task Force IV: Transparency
Chair: Roger L. Hagengruber

Kenneth E. Baker
U.S. Department of Energy

Donald D. Cobb
Los Alamos National Laboratory

George Eccleston
Los Alamos National Laboratory

R. Charles Gentry
Independent Consultant

Yafei He
Embassy of the People's Republic of China

Edward R. Johnson
JAI Corporation

Ben Sanders
Programme for Promoting Nuclear Non-Proliferation

Thomas Sellers
Sandia National Laboratories

Alexander Stoliarov
Bechtel

Elizabeth Turpen
Office of Senator Pete V. Domenici

Task Force V: U.S. Domestic Infrastructure and the Emerging Nuclear Era
Chair: John J. Taylor

Thomas E. Blejwas
Sandia National Laboratories

Paul T. Dickman
U.S. Department of Energy

Dale E. Klein
University of Texas System

Peter Lyons
Office of Senator Pete V. Domenici

William D. Magwood IV
U.S. Department of Energy

Robert Mahar
Westinghouse Savannah River Company

Robert E. Meadors
Westinghouse Savannah River Company

John Stamos
U.S. Department of Energy

Kristine Svinicki
Office of Senator Larry E. Craig

Wayne Willis
General Atomics

Coordinating Committee
Chair: Matthew Bunn

Harold D. Bengelsdorf
Bengelsdorf, McGoldrick and Associates

Paul T. Dickman
U.S. Department of Energy

Marv Fertel
Nuclear Energy Institute

Fred McGoldrick
Bengelsdorf, McGoldrick and Associates

Daniel Poneman
Hogan & Hartson, L.L.P.

Thomas L. Sanders
Sandia National Laboratories

Kristine Svinicki
Office of Senator Larry E. Craig

Elizabeth Turpen
Office of Senator Pete V. Domenici

Executive Summary

New Leadership to Reduce the Nuclear Threat: The Vision and the Immediate Priorities

Despite the end of the Cold War, nuclear weapons continue to pose the most devastating security threat to Americans. Although the risk of a nuclear war destroying civilization has virtually disappeared, the risk that a single nuclear weapon might be used to destroy a major city has increased, particularly given the erosion of control over nuclear material with the collapse of the Soviet Union. Nothing could be more central to international security than ensuring that the essential ingredients of nuclear weapons do not fall into the hands of terrorists or proliferant states. Effective controls over nuclear warheads and the nuclear materials needed to make them are essential to the future of the entire global effort to reduce nuclear arms and stem their spread. At the same time, ensuring protection of public health and the environment in the management of all nuclear materials—from nuclear weapons to nuclear wastes—remains a critical priority. Appropriate management of both safety and security worldwide will be essential to maintaining nuclear fission as an expandable option for supplying the world's greenhouse-constrained energy needs in the twenty-first century.

The vision of global nuclear materials management is of a world in which all nuclear materials are safe, secure, and accounted for, from cradle to grave, with sufficient transparency to assure the world that this is the case. That is a daunting goal, which must be approached step by step, within a well-defined strategic framework. The Senior Policy Panel of this project has identified two key areas where the need for action is particularly urgent:

The Eroding Controls in the Former Soviet Union

The combination of insecure, oversized nuclear stockpiles and an underfunded nuclear complex in the former Soviet Union (FSU), managed with little or no international transparency, poses a severe threat to U.S. and international security. The possibility that the essential ingredients of nuclear weapons could fall into the hands of terrorists and proliferating states is all too real, and immediate actions are needed to reduce this threat to the security of the United States and the world. The recent Expanded Threat Reduction Initiative is only the beginning of what needs to be done. We have developed a recommended action plan of urgently needed steps in this area.

A Withering Foundation for U.S. Leadership

Judged by any of a broad range of criteria, the infrastructure of U.S. leadership in nuclear technologies has greatly weakened over the last two decades. U.S. nuclear research and development (R&D) is dwarfed by R&D under way in other nations, the cadre of experienced personnel is dwindling, and nuclear engineering departments at U.S. universities are shrinking. The United States has virtually disengaged from international discussions and cooperation on the future of the nuclear fuel cycle. If the United States can no longer credibly claim a leadership role in nuclear technology or is seen as having no interest in the future of nuclear energy, its ability to lead in nonproliferation could be substantially undermined. Here, too, immediate action is needed to rebuild the R&D program, a cadre of experts, R&D facilities, and materials infrastructure to help provide the foundation for global leadership. Here, too, we have outlined series of steps for near-term action.

These are category 1 priorities for U.S. national security policy. As Senator Joseph R. Biden Jr. (D-Del.) recently put it: "The war against these 'loose nukes' and 'brain drain' threats is as important as any war in our history…it is a war that the United States dares not lose." Major programs are under way to address these threats, and a small beginning is being made on rebuilding the U.S. nuclear technology infrastructure, but much more remains to be done. For the United States to address these issues successfully will require a sea change in the level of sustained, high-level leadership devoted to them—including the personal involvement of the president and the vice president. The five task forces of this study, some of which are described below, have outlined a rich menu of approaches that could quickly and demonstrably reduce the risks we face and increase the potential for continued U.S. leadership. The time for action is now. The costs and risks of failure to act are far higher than the costs of timely action to prevent disaster.

Reducing the Risk of Nuclear Theft and Laying a Basis for Irreversible Reductions

The Problem

Theft of just a few kilograms of plutonium or highly enriched uranium (HEU)—the essential ingredients of nuclear weapons—could allow a rogue state or terrorist group to acquire a nuclear capability, posing a severe threat to the international community. This risk is a global problem requiring global solutions. Today, however, the problem is most acute in Russia, where the world's largest stockpile of weapons-usable material resides—more than 1,000 tons of HEU and plutonium, roughly half in weapons and the rest in a wide variety of forms distributed over some 300 buildings at more than 50 sites. Moreover, none of this material, whether civil or military, is under international safeguards.

The collapse of the Soviet Union and the ensuing economic and political turmoil in the former Soviet states dramatically weakened controls over nuclear materials there. A nuclear security system built for a single state with a closed society with closed borders and well-paid nuclear workers has splintered among multiple states with open societies, open borders, and desperate, unpaid nuclear workers. Nuclear guards reportedly go unpaid for months at a time and leave their posts to forage for food; nuclear security systems go unmaintained or even unused for lack of funds; electricity that provides the lifeblood of nuclear alarm and monitoring systems is shut off for nonpayment of bills; and in several documented cases, kilogram quantities of weapons-usable nuclear material are stolen. In Russia, all this is taking place in a nation with a collapsing economy, rampant organized crime, and persistent corruption at many levels of government, where virtually every commodity is for sale if the price is right. Even the Russian minister of atomic energy has acknowledged that the reduction in Russia's ability to control nuclear materials has been "immeasurable." The Central Intelligence Agency has gone further, warning that the risk that potential bomb materials could fall into the hands of terrorists or proliferant states is higher than ever before.

At the same time, Russia has not yet downsized its Cold War nuclear complex. The complex is both oversized (which poses a risk to the United States because Russia could rapidly return to producing thousands of nuclear weapons a year, should circumstances change) and underfunded (which poses risks because the desperation of poorly paid workers creates incentives for sale of nuclear materials or nuclear knowledge, particularly when the barriers to proliferation are also underfunded). The Russian nuclear complex still includes 10 entire nuclear cities—cities built only to produce nuclear weapons and their ingredients—fenced off from the outside world and guarded by armed troops—but with a vastly reduced mission and collapsing budgets. It is in both U.S. and Russian interest to shrink this huge complex to a more appropriate and sustainable level, eliminating excess weapons production capacity while providing alternative employment for the nuclear weapons scientists and technicians who are no longer needed for stewardship of Russia's stockpile.

Finally, Russia and the United States both still maintain very large stockpiles of nuclear weapons and the plutonium and HEU needed to make them. Just one of the Russian nuclear cities holds more plutonium and HEU than the arsenals of Britain, France, and China combined. These vast stockpiles are managed with very little of the transparency that would be needed to build confidence that they are safe and secure or to provide the foundation for deep, transparent, and irreversible nuclear arms reductions. (Transparency is also critical to ensuring that U.S. assistance is spent appropriately, and a number of steps toward achieving that goal have been successfully implemented.)

The Programs

The United States has put in place a broad range of programs costing hundreds of millions of dollars annually to help the former Soviet states in addressing these threats. These are briefly described in the report of Task Force I. Important efforts are under way at the Department of Defense (where the original Nunn–Lugar Cooperative Threat Reduction program resides), the Department of Energy (DOE), the Department of State, the Customs Service, the Federal Bureau of Investigation, the Nuclear Regulatory Commission, and elsewhere. These programs represent some of the most cost-effective investments in U.S. national security found anywhere in the U.S. budget and deserve strong and continuing support. Other nations have also contributed to the effort, although on a much smaller scale, and need to do more. President Bill Clinton's proposal for an Expanded Threat Reduction Initiative recognized the importance of these threats and called for additional funds in some areas. Unfortunately, however, with respect to addressing the "loose nukes" threat, this proposal in essence only continues flat funding for programs that once had been planned to decline, rather than stepping up the level of U.S. efforts or launching major new initiatives. As outlined in the task force reports (particularly the report of Task Force I), enormous and urgent needs remain for additional efforts to address this security problem facing the United States, Russia, and the rest of the world.

The Partnership

Nuclear insecurity problems in the FSU can only be successfully addressed with a true spirit of partnership with the former Soviet states and their experts. Attempts to dictate specific Western approaches or impose solutions will inevitably fail in the long run and will undermine the prospect for intensified cooperation. Ultimately, if an expanded agenda of nuclear security cooperation is to be successful, experts from the former Soviet states will have to play central roles in its design and implementation.

The Constraints

The policy problem is to identify a set of actions that would make a major contribution to reducing these threats while, at the same time, being capable of gaining political support in the United States and in the former Soviet states, particularly Russia. Each of Task Forces I–IV describes different aspects of the difficult constraints that limit the prospects for expanding particular U.S.-Russian cooperative nuclear security efforts. Sour U.S.-Russian political relations in the aftermath of the air campaign against Yugoslavia may constrain what can be done in some areas—but this political reality also highlights the importance of new efforts to reinvigorate genuine cooperation in areas that serve both U.S. and Russian interests. Problems from widespread corruption to Russian efforts to tax U.S. assistance make it critical to put high priority on ensuring that U.S. taxpayer dollars are spent appropriately. Secrecy concerns—some legitimate, some overdrawn—limit access to information and facilities, restraining cooperation. (Despite the necessity of improving security for U.S. secrets in the wake of recent revelations of Chinese espionage, it is critical not to impose constraints that would limit cooperation between U.S. lab personnel and their international counterparts. Cooperation, such as ensuring that bomb material does not fall into the wrong hands, is critical to U.S. security objectives.) Commercial approaches to converting the Russian nuclear and defense infrastructure are hobbled by a combination of the problems besetting the Russian economy as a whole (including particularly daunting tax and legal obstacles to successful investment and business operation), the problems that have limited the success of defense conversion even in thriving market economies, and problems unique to Russia's nuclear complex, particularly secrecy and limits on access to the nuclear cities. It is clear that government investment will be needed to help overcome these obstacles and leverage private sector funds.

The Necessary Next Steps

Despite these obstacles, the panel is convinced that the time is right for major new U.S. and international initiatives to reduce these critical threats to U.S. and international security. It is simply unacceptable to continue a situation in which lack of sufficient funding and senior leadership attention on the U.S. side are among the major factors preventing faster and more effective actions to reduce these serious security threats. The time has come to outline what it would take to reduce these threats as fast as it is realistically practicable to do so. Such a new program of fissile material threat reduction should focus on efforts to buy, consolidate, secure, monitor, and reduce weapons-usable nuclear material stockpiles; shrink the Russian nuclear complex; and ensure sustainable security for the future.

BUY. Buying Russian HEU—which, when blended to low-enriched uranium, is both proliferation resistant and commercially valuable—is the closest thing yet devised to a "silver bullet" for addressing the huge, complex, and multifaceted problems of nuclear security in the FSU. The U.S.-Russian HEU purchase agreement (covering 500 tons of HEU from dismantled weapons over 20 years) is converting thousands of bombs' worth of weapons material to peaceful reactor fuel, providing a financial incentive for warhead dismantlement, giving the United States unprecedented transparency at several major Russian military nuclear facilities, and providing hundreds of millions of dollars a year to stabilize the desperate Russian nuclear complex, all primarily on a commercial basis, at minimal cost to the U.S. taxpayer. The panel commends those who worked successfully to achieve the recent government-to-government and commercial agreements to get this deal back on track. The U.S. government must place very high priority on ensuring that this deal continues to move forward in the future—an issue that is likely to arise again soon, as the current contract for purchases of the enrichment component comes up for renegotiation.

In addition to preserving the achievements of the past, however, the time has come to buy more HEU, and faster, in three ways (each is described in the report of Task Force I):

- Offer to buy substantial additional quantities of HEU, with a portion of the proceeds designated in the contract to go to an auditable fund to pay for nuclear security in Russia—ensuring that nuclear guards and workers are paid, and security systems operated, maintained, and improved. As an initial step, this might involve an additional 50 tons, at a cost of roughly $1 billion, paid with government money, to be held off the market as a uranium reserve, as is being done with much of DOE's uranium stockpile now (so as not to disturb the commercial arrangements for the original deal).
- Seek to buy not only HEU from dismantled warheads, but the small, vulnerable HEU stockpiles located at small research facilities that can no longer afford to guard it—while providing assistance to close these facilities or help the research

facilities pursue research that does not require HEU. Because the amounts are small (only a few tons in total), the cost would be small as well.

- Offer to provide the needed capital investment and financial incentives to make it possible for Russia to blend all its excess HEU to non-weapons-usable form within the next few years—perhaps blending it first to an intermediate level such as 19 percent U-235 before later final blending and purification for sale. Rapid blend-down could address the urgent security risks while the material is released into the commercial market at the previously agreed pace.

CONSOLIDATE. New U.S.-Russian cooperative efforts to consolidate material at fewer locations should be accelerated and expanded. Consolidation is a critical priority, allowing greater security at the remaining locations to be achieved at lower long-term cost (although significant initial investment is likely to be required to make consolidation happen). One approach that ought to be considered is providing financial incentives for depositing the majority of Russia's plutonium and HEU in one or more internationally monitored storage facilities (possibly with a similar facility being established in the United States for reciprocity). Facilities that already exist or are under construction (particularly the Mayak storage facility being built for fissile materials from dismantled weapons) could be used for such a program.

SECURE. No matter how much material is purchased or transformed into non-weapons-usable forms and how much the remainder is consolidated, a substantial number of facilities in the FSU with nuclear weapons, plutonium, and HEU will remain. It is critical to ensure that all of these stockpiles are secure and accounted for as rapidly as practicable. The current cooperative effort to improve material protection, control, and accounting (MPC&A) for weapons-usable nuclear material in the FSU is making good progress and deserves strong support (although a variety of issues continue to arise that constrain cooperation), as do the much smaller efforts of several other countries, the International Atomic Energy Agency (IAEA), and the European Union. Nevertheless, the panel is convinced that substantially more funding and more leadership resources are required to improve security and accounting for this material as rapidly as it would be practicable to do so, while simultaneously moving quickly on consolidation and on putting in place a truly sustainable security system for the future. DOE should develop and propose a program designed to reduce the urgent proliferation threats posed by insecure nuclear material as quickly as realistically possible, and Congress should give that program its support.

MONITOR. As described in detail in the report of Task Force IV, increased transparency in the management of nuclear warheads and materials—with continued protection of legitimate nuclear secrets—is critical to a variety of objectives, from building confidence that warhead and fissile material stockpiles are being safely and securely managed to providing the foundation for deep reductions in stockpile size. Transparency is also critical to ensure that U.S. and international assistance is being appropriately spent. Unfortunately, a range of factors are likely to make progress on the possible transparency initiatives outlined in the reports of Task Force IV and Task Force I extraordinarily difficult to achieve in the near term.

These factors include strained U.S.-Russian political relations, a reinvigorated Russian security service, the distractions of upcoming elections in both countries, and a renewed U.S. focus on protecting nuclear secrets that is eclipsing the potential security benefits of nuclear openness.

This proposed effort should by no means be abandoned, however, despite the recognition that transparency in nuclear weapons has limitations. The United States should, first, take steps to make transparency progress in Russia's own interest, through offering strategic and financial incentives. Second, it will probably be necessary to begin with small steps, to rebuild the foundation for trust over time. For example, while the panel judges it quite unlikely that a complete warhead transparency regime could be prepared in time to be part of a Strategic Arms Reduction Talks (START) III treaty before the current U.S. and Russian presidents leave office, some initial exchanges and demonstrations of technologies and procedures might well be possible in that time frame if coupled with a broader package of incentives.

Reduce. The vast stockpiles of plutonium and HEU built up over decades of Cold War are far larger than needed in the post–Cold War era and must be reduced—converted to forms much less usable in nuclear weapons. These huge stockpiles will pose serious security risks as long as they remain in readily weapons-usable form. The former Russian minister of atomic energy, Viktor Mikhailov, once said, "Real disarmament is possible only if the accumulated huge stocks of weapons-grade uranium and plutonium are destroyed." As arms reductions proceed, these stockpiles should be reduced in parallel to roughly equivalent levels in the United States and Russia. These levels should be suitable to support whatever agreed warhead levels remain but not large enough to permit a rapid return to Cold War levels of armament.

As noted earlier, the HEU stockpile can readily be reduced by blending with other forms of uranium and sold on the commercial market. Excess plutonium poses far more difficult obstacles, and it will inevitably be many years before all the excess plutonium has been transformed into forms that are no longer directly usable in nuclear weapons. Nevertheless, the panel concurs with numerous previous reports on this issue—including the March 1998 CSIS panel report, *Disposing of Weapons-Grade Plutonium*—that reducing excess plutonium stockpiles as rapidly as practicable is a high priority. Plutonium disposition is in the unusual position of being a long-term issue requiring urgent action—in part because the U.S. Congress has made clear that it will not support requests for funding the construction of plutonium disposition facilities in the United States in the absence of substantial progress toward a clear commitment that Russian plutonium stockpiles will be reduced in parallel. While considerable progress is being made in U.S.-Russian discussions regarding such commitments, the fundamental issue of who will pay the more than $1 billion cost of disposition of the Russian plutonium remains unsolved. The United States, at the initiative of Senator Pete V. Domenici (R-N.Mex.), made a substantial step toward resolving this problem by appropriating $200 million for a first installment in fiscal year (FY) 1999.

In the panel's judgment, the time has come for the United States to either commit to pay the full cost itself or reach agreement with other leading nuclear countries on an equitable financing scheme that will allow the program to move forward rapidly. Compared with the national security risk, the required funds are insignificant. First priority should be placed on ensuring that the material is securely stored and accounted for, converted to unclassified forms, and placed under bilateral and/or international monitoring.

SHRINK. More must be done to shrink the enormous Russian nuclear complex to a sustainable size suitable for its post–Cold War missions and provide appropriate civilian work for the facilities, scientists, and workers who are no longer needed. The United States and the international community can help with dismantling or converting facilities once used for nuclear weapons missions and providing support for a variety of targeted efforts to create new jobs for excess workers. Russia itself, of course, must shoulder the burden of financing the maintenance of a smaller, safe, and secure weapons complex to provide stewardship for the reduced nuclear weapons stockpile Russia will surely retain. Task Force III outlines the obstacles to success in building a sustainable commercial future for these facilities but also provides a number of valuable suggestions for steps that could be taken. A substantial increase in investment, from both the United States and other leading industrialized countries, is likely to be necessary if these root causes of the nuclear insecurity problems in the FSU are to be successfully addressed.

SUSTAIN. U.S. and international financial assistance for nuclear security in the FSU will not, and should not, last forever. Ultimately, the former Soviet states will bear the full burden of ensuring safety and security for all their nuclear materials themselves. U.S. and international efforts to work with the former Soviet states to build these states' own capacity to provide sustainable security over the long run must be radically increased. In particular, given the grave continuing economic difficulties in many of these states, intensive efforts are needed to identify new revenue streams that could support nuclear security activities, both immediately and after international assistance ends—from MPC&A to disposition of excess plutonium. One possibility, addressed in detail in the report of Task Force II, is to use revenue from the establishment of an international facility for storage or disposal of spent nuclear fuel, in Russia or elsewhere. Task Force II clearly outlines the substantial obstacles still facing such proposals—but also describes some of the incentives for moving forward with such an effort. It suggests that the United States should be prepared to outline the criteria such concepts would have to meet to win U.S. support (which will be essential for such facilities to receive any of the large fraction of the world's spent fuel over which the United States has consent rights). In addition to offering the potential for substantial revenue to support nonproliferation and cleanup in the FSU, an international storage or disposal facility would offer substantial additional benefits as well, including offering a proliferation-resistant way to manage growing global inventories for spent fuel.

The Role of the Private Sector

Commercial industry—particularly the nuclear industry—has an essential part to play in addressing many of these issues, bringing expertise, experience, and entrepreneurial energy to the table. But as Task Force III points out, the private sector cannot do the job alone: the obstacles and risks to commercial success in many of these areas are too great, and the avenues for potential profit too few. Government funding will be necessary to fund some of these efforts in their entirety, and others in part. Government funding should be targeted to areas that must remain in government control and those areas where government support can overcome obstacles and manage risk, thus leveraging far larger flows of private capital. The private sector has a particularly critical role to play in the effort to provide alternative jobs for Russia's excess nuclear workers and facilities and in the management of the vast excess HEU and plutonium stockpiles in the United States and Russia.

The Role of Other Leading States and Organizations in the International Community

Nuclear insecurity is not just a U.S. and Russian problem. It affects the entire international community. To date, the United States has done much more to provide funding to reduce these nuclear security threats than have other major industrialized nations. The panel believes that the time has come for leading developed states in Europe and Asia to increase substantially their contributions in these areas. Japan's recent announcement of a new $200 million contribution to submarine decommissioning and plutonium disposition is a welcome first step in this direction. At the same time, it would be tragic if the possibility of larger contributions from other countries were used in the United States as an argument against providing U.S. funding for these extraordinarily cost-effective investments in U.S. national security.

The Need for Leadership

Reducing the nuclear threat before catastrophe strikes will require energetic and visionary leadership, pulling together a broad range of critically important initiatives into an integrated effort. A sea change in the level of sustained leadership from the highest levels of the U.S. government—including the president and the vice president—is needed. The panel recommends designating a senior, full-time point person for this task, with direct access to the president, as was recently done for reviewing U.S. policy toward North Korea. Preventing nuclear material from falling into the hands of states like North Korea or Iraq is certainly no less critical to U.S. security; indeed, the entire global effort to prevent the spread of nuclear weapons depends on it.

Beyond the Former Soviet Union: A World of Issues

As the name implies, the vision of global nuclear materials management is not limited to the states of the former Soviet Union. Although this report has focused primarily on the urgent needs for improved nuclear materials management in those states, there are key issues of nuclear materials management the world over that affect the safety and security of the international community, and whose mismanagement could undermine the future prospects for nuclear energy as an important potential contributor to the world's energy needs in the twenty-first century. These key issues include management of spent nuclear fuel from nuclear reactors, safe disposal of high-level wastes, efforts to ensure and improve the effectiveness of international safeguards, and steps to reduce the security risks posed by plutonium and HEU in the rest of the world outside the FSU (symbolized by concern during the 1999 air campaign in the Balkans over the fate of the weapons-usable HEU located at a research reactor in Yugoslavia). Greatly increased international cooperation—ideally with a reinvigorated foundation for U.S. leadership—will be needed to address these problems. Space does not allow a full exploration of these issues, but the panel believes that several key action items related to the global management of nuclear material should be highlighted:

- The IAEA plays an absolutely central role in safeguarding nuclear material and working with states to improve its management worldwide. The IAEA's workload has increased dramatically in recent years and will increase further in the future if efforts such as the negotiation of a fissile cutoff treaty and IAEA verification of nuclear material rendered excess by disarmament bear fruit. Having been limited to a zero-real-growth budget since the mid-1980s, the IAEA urgently needs additional funding, and the United States and other major nuclear powers should redouble their efforts to provide it.

- Spent fuel storage facilities around the world are filling up; there is a compelling need for new leadership to help establish additional spent fuel storage facilities and increase the pace of progress toward establishment of permanent repositories. These could include both national facilities and regional or international facilities designed to serve the needs of multiple states. Resolution of the long-running conflict between the U.S. government and U.S. nuclear utilities over responsibility and approaches for spent fuel management is essential to help repair the seriously damaged U.S. credibility on this issue.

- As the President's Committee of Advisors on Science and Technology has recently recommended, the United States should undertake a new international cooperative initiative to promote safe and proliferation-resistant spent fuel interim storage in both national and international facilities. The United States should be prepared to help make the case for interim storage of spent nuclear fuel as a safe and cost-effective near-term approach to spent fuel management, as well as provide technology and funding in some limited cases. The United States must also move forward expeditiously on completing the scientific studies and reaching a presidential decision on the suitability of the Yucca Mountain site for a permanent repository.

- As described in the report of Task Force III, increased nuclear transparency is needed not just between the United States and Russia but for the international community as a whole. In particular, if nuclear energy is to rebuild the level of government, utility, and public support that would be required for it to grow substantially in the twenty-first century, there will be a central need to provide sufficient transparency to assure the public that its concerns are being effectively addressed, and that nuclear materials and the nuclear enterprises are being safely and securely managed.

- New steps are needed to ensure that plutonium and HEU worldwide, not just in the FSU, are secure and accounted for. Expanded international cooperation is called for to ensure that states are meeting international standards that effectively secure and account for all plutonium and HEU worldwide.

Rebuilding the Foundation for U.S. Technological and Nonproliferation Leadership

U.S. leadership on nonproliferation and safety issues (particularly as they relate to both the government infrastructure and civilian nuclear energy) is fundamentally linked to the strength of its technical foundation, to the perception of the commitment of the U.S. government to maintaining a nuclear power option for the future, and to the policy positions taken by the United States. Unfortunately, at the cusp of the twenty-first century and of a new nuclear era, with critical nuclear security issues around the globe crying out to be addressed, the United States has allowed the essential technical foundations of its leadership in nuclear nonproliferation and safety to atrophy and has greatly decreased its participation in international cooperation on nuclear energy and its fuel cycle.

Although the United States still has the largest number of operating nuclear reactors in the world, and reactors with advanced U.S. designs are being built and operated in Asia, no new nuclear reactor has been ordered in the United States for decades. U.S. government-sponsored R&D on civilian nuclear energy fell to zero in FY 1998 and is still at a historic low. The first generation of nuclear technologists is past retirement age; the number of students in U.S. nuclear engineering programs is plummeting, and 70 percent of those who do receive nuclear engineering graduate degrees in the United States are foreign. The United States is doing very little to help develop improved safety, proliferation resistance, waste management, and economics for the nuclear power of the future; and it has lost the lead in many areas of nuclear technology, notably test facility capability, nuclear plant fabrication and construction, and certain aspects of the nuclear fuel cycle. The U.S. nuclear industry has been left alone to compete in a rate-deregulated market, without any credit for the atmospheric emissions its energy generation helps to avoid. With the end of the Cold War, DOE's infrastructure for managing nuclear materials has shrunk, narrowing the options for dealing with the Cold War's nuclear residues. Although current efforts to rebuild the U.S. R&D effort and technical base should be commended, they represent far less than what is required. The effective U.S. withdrawal from international discussions and R&D on long-term approaches to the fuel cycle in recent years has been particularly damaging to U.S. nonproliferation leadership.

The United States must plan for an international future that will have fewer nuclear weapons but more nuclear waste and more excess defense materials and that may see larger and more widespread use of nuclear energy. Halfway through the first nuclear century, we are at a time of enormous challenges, opportunities, and transition. How the United States responds will determine how it is perceived by other countries. The time has come for a clear statement from the highest levels of the U.S. government that the United States believes it is important to maintain the nuclear option as a potentially critical contributor to meeting the world's energy needs, which will be constrained in the twenty-first century by potential controls on emissions of both traditional pollutants and greenhouse gases. At the same time, immediate actions are needed to rebuild the technical underpinning of U.S. nonproliferation and safety leadership. First is expanding the U.S. nuclear R&D program, so that the United States is involved in and understands the technologies whose safety and nonproliferation impacts it is attempting to influence, and so that it can be at the cutting edge in developing the safe and proliferation-resistant technologies of the future. Second is rebuilding the cadre of nuclear experts who understand these technologies and experts in proliferation and safeguards who can help control such technologies. Third is providing the facilities required for these experts to carry out an effective R&D program and for safe and proliferation-resistant management of U.S. nuclear materials; such facilities are critical for maintaining a viable nuclear industry. Specific recommendations are:

- Funding for nuclear R&D should be substantially increased and focused on the critical safety, nonproliferation, waste management, and cost issues that have constrained nuclear power's growth to date;
- Within that R&D portfolio, specific steps should be taken to reinvigorate the nation's nuclear engineering departments and attract new students to the field (including a new generation of proliferation and safeguards experts), and to rebuild the infrastructure of facilities needed for such R&D;
- Steps should ensure that the United States retains the infrastructure needed for an effective R&D program and for effective management of its nuclear materials;
- A new initiative should be undertaken for international cooperation in such R&D, as recently proposed by the President's Committee of Advisors on Science and Technology, in recognition of the potentially important world role nuclear power could have in addressing global warming;
- The United States should reengage in international discussions and R&D on safe and proliferation-resistant approaches to the fuel cycle. Finding ways to better utilize limited nuclear resources and ensure adequate fuel supplies for the long term would range from conceptual studies of more proliferation-resistant recycling systems to explorations of whether recovery of uranium from unconventional resources such as seawater may be viable;
- Operators of nuclear facilities should place the highest priority on the safe operations of their plants and should work to reduce incidents that could undermine public trust in nuclear power;
- Further steps should be taken toward a more risk-informed, performance-based nuclear safety regulatory process, and wasteful overlaps in regulatory jurisdictions should be eliminated;
- Regulators should redouble their efforts to ensure that license renewals and reviews of changes in ownership of nuclear plants are handled expeditiously;

- Final agreements should be reached on responsibility, financing, and approaches for storage of commercial spent fuel, pending availability of a permanent repository. Efforts to resolve the safety issues and address public concerns so as to allow a permanent repository to open should also be redoubled;
- Appropriate public policy recognition should be given to nuclear power's ability to generate electricity with zero emissions of carbon dioxide, sulfur and nitrogen oxides, and particulates. Consideration should be given to allocating tradable emission permits equally among all generators of electricity—fossil, nuclear, and renewable—on the basis of power output, and thus giving nuclear and renewable energy sources credit for the emissions they avoid.

A Call to Action

The world simply cannot afford delay in addressing the urgent security hazards posed by nuclear insecurity in the FSU. There is little remaining margin for continued decay of the U.S. nuclear infrastructure if the United States is to be technically credible in nonproliferation leadership in the twenty-first century. The opportunities are there; an investment of a few billion dollars, properly applied, could dramatically reduce the risks the world now faces. The fundamental requirement is leadership. The time to act is now—*before* a catastrophe occurs.

Part One

CHAPTER 1

Funding Nuclear Security

Task Force I

WITH MORE RESOURCES, WHAT COULD BE DONE TO SECURE NUCLEAR WARHEADS AND FISSILE MATERIALS?

Today insecure and oversized nuclear weapons and materials stockpiles in the former Soviet Union (FSU), managed with little international transparency, coupled with an oversized and underfunded nuclear complex there, pose severe threats to international security. Nothing could be more central to U.S., Russian, and world security than ensuring that the essential ingredients of nuclear weapons do not fall into the wrong hands. Moreover, measures to control nuclear warheads themselves and the fissile materials needed to make them are also essential to achieving deep, transparent, and irreversible nuclear arms reductions. Although secure management of nuclear material is a global issue requiring global solutions, the issues are currently most acute in the FSU, where the world's largest stockpiles of weapons-usable material reside and where post–Cold War political and economic upheavals have severely undermined previous security arrangements.

These are category one priorities for U.S. national security policy. As Senator Joseph R. Biden Jr. (D-Del.) has recently put it: "The war against these 'loose nukes' and 'brain drain' threats is as important as any war in our history...it is a war that the United States dares not lose."[1] To address these issues, the U.S. government has put in place a broad range of programs costing hundreds of millions of dollars a year that are doing excellent work and deserve strong support (see appendix A of this chapter, "Current Programs," page 17).[2] Yet the level of investment being devoted to these security hazards is tiny in comparison with what the United States has been accustomed to spend, and continues to spend, to provide defenses against military threats to its national security. (See appendix B of this chapter, "Sizing the Problem," on page 19, for a discussion of problems posed by Russia's nuclear infrastructure.)

---

1. Senator Joseph R. Biden Jr., "Maintaining the Proliferation Fight in the Former Soviet Union," *Arms Control Today (*March 1999).

2. Note that the focus of this chapter is on the management of nuclear weapons and weapons-usable materials, and therefore a wide range of other worthy cooperative threat reduction efforts are not addressed—from dismantling missiles to destroying chemical weapons, from improving export controls to reemploying biological warfare experts.

The job of this task force was to examine whether there were opportunities to reduce these security risks more rapidly and effectively if more money were spent—while maintaining a central focus on ensuring that the money is spent appropriately for its intended purposes. The short answer is yes.

The task force believes that although there are many nonmonetary obstacles to U.S.-Russian nuclear security cooperation (see appendix C of this chapter, "Obstacles to Nuclear Security Cooperation," page 21), there are additional steps that could be taken to reduce these risks if additional resources were applied to the task by the U.S. government and other governments. Indeed, in addition to the direct security benefits, a significantly expanded program of nuclear security cooperation could play an important role in improving U.S.-Russian relations in the security sphere. The measures called for in the Expanded Threat Reduction Initiative the Clinton administration has proposed are only the beginning of what needs to be done; indeed, for programs to improve controls over nuclear materials, the Clinton administration initiative calls only for continued flat funding (such funding had previously been scheduled to decline) not major new funding or initiatives. All of the suggestions outlined below are in addition to the programs called for in the Expanded Threat Reduction Initiative.

The specific suggestions in this task force report were prepared largely by Americans. Nonetheless, if one lesson comes through loud and clear from the experience of programs in these areas, it is that success can only be achieved in true partnership with Russian experts, with their perspectives taken fully into account. Ultimately, if an expanded agenda of nuclear security cooperation is to be successful, Russian experts will have to play central roles in its design and implementation.

The United States has two basic objectives in its cooperative programs related to the control of nuclear warheads and fissile material: to reduce the risk of nuclear proliferation; and to achieve deep, transparent, and irreversible nuclear arms reductions. The myriad cooperative warhead and fissile material programs designed to pursue these goals can be divided into five basic categories, enumerated below with a description of the basic objectives for each of the five:

- **Preventing theft and smuggling:** Ensure that all weapons-usable nuclear material is secure and accounted for and provide a second line of defense with measures to interdict smuggling of stolen nuclear material;
- **Stabilizing nuclear custodians:** Provide sustainable civilian jobs for excess nuclear workers, shrink the Russian nuclear weapons complex, and ensure that workers and guards who continue to have access to weapons-related information and materials are trained, paid, housed, and the like;
- **Monitoring stockpiles and reductions:** Build, through a step-by-step approach, a transparency regime that can provide confidence that warhead and fissile materials stockpiles are being reduced to low, agreed levels and are secure and accounted for;
- **Ending production of fissile material:** Verifiably end production of highly enriched uranium (HEU) and separated plutonium for weapons;

- **Reducing stockpiles:** Transform excess HEU and plutonium into forms that are no longer usable in nuclear weapons, leaving only enough in military stocks to support low agreed warhead levels and naval programs.

What follows below is a listing of steps to address each of these five objectives that could be enabled by additional resources. These are divided into three broad categories: *modest steps* (initiatives with price tags in the range of millions to a few tens of millions), *significant strides* (initiatives in the range of a few hundreds of millions over several years), and *great leaps* (initiatives that would cost billions of dollars, coming close to matching the level of resources to the importance of these threats to U.S. and international security).

In addition, in appendix D of this chapter, the task force describes several concepts that have been proposed that could provide additional revenue streams to address these proliferation threats in the FSU (see appendix D of this chapter, "Generating New Revenue for Nuclear Security," page 23). Both the suggested initiatives and the cost estimates associated with them are intended to be illustrative, not definitive, and this list of possibilities is by no means complete.

Not every member of the task force agrees with every detail; but the task force is in agreement that despite the recent downturn in U.S.-Russian relations and the myriad other obstacles facing such cooperation, a wide range of possibilities exists for further action to address the post–Cold War security risks posed by the nuclear stockpiles and the nuclear complex in the FSU that could be enabled through energetic leadership and the application of additional financial resources.

Essentially all of the recommended steps could be taken within the context of existing cooperative programs, ranging from the Nunn–Lugar Cooperative Threat Reduction program at the Department of Defense (DOD) to efforts such as plutonium disposition and material protection, control, and accounting at the Department of Energy (DOE). The task force has not indicated which department or program should pay for each of the recommended steps, preferring to allow the allocation of responsibilities within the administration to continue to evolve—as long as an appropriate level of urgency and overall strategic planning is brought to the task (an issue we describe at the end of this report). Table 1.1 outlines our suggested agenda for action under the three categories: modest steps, significant strides, and great leaps.

Modest Steps

Preventing Theft and Smuggling

Expand MPC&A. Increased funding for the cooperative material protection, control, and accounting (MPC&A) program (beyond the $140 million allocated in FY 1999 and the $145 million requested for FY 2000) would make it possible to allocate significant funding to consolidating vulnerable nuclear material stockpiles at fewer locations (as the Russian Ministry of Atomic Energy [MINATOM] has now agreed to do) and to measures improving the sustainability of security and accounting upgrades, without reducing the pace of installing security and accounting

Table 1.1 Agenda for Action

Modest Steps	
	Expand the cooperative material protection, control, and accounting (MPC&A) program
	Purchase small, vulnerable HEU stockpiles
	Expand assistance for interdicting nuclear smuggling
	Target a portion of economic assistance to Russia to the nuclear cities
	Finance data exchanges on nuclear material stockpiles
	Finance international monitoring of excess fissile material
	Confirm nonproduction of HEU
	Examine feasibility and cost of rapid HEU blend-down
Significant Strides	
	Expand MPC&A to a level not constrained by funding
	Expand nuclear-cities initiative to support a broad range of sustainable employment
	Provide financial assistance for transparent warhead dismantlement
	Finance verification of a fissile cutoff in Russia
	Finance rapid blend-down of excess HEU
	Finance construction of necessary facilities for plutonium disposition
Great Leaps	
	Provide incentives to consolidate HEU/plutonium in internationally guarded facilities
	Establish a comprehensive program to downsize the Russian nuclear complex and provide alternative employment
	Propose and finance wide-ranging reciprocal warhead- and materials-monitoring regime
	Purchase additional quantities of excess HEU
	Purchase Russian excess plutonium

upgrades and providing training that was achieved in FY 1998 and FY 1999. Measures to build sustainability and promote the growth of a modern safeguards culture are particularly critical to ensure that U.S. assistance actually results in lasting security and accounting improvements and that security and accounting equipment provided with U.S. assistance is actually operated, maintained, and improved over time. Additional assistance for material accountability is also critical, placing first priority on those measures that can be accomplished quickly, including identifying, tagging, and sealing all the items containing weapons-usable material. New attention to types of material that have previously been neglected—such as HEU spent fuels for naval and breeder reactors, which also pose proliferation risks—is also needed.

This would require added funding of approximately $20 million–$165 million per year.

Purchase small, vulnerable HEU stockpiles. There are many small nuclear research facilities in the FSU that have significant quantities of HEU but no longer have the financial resources to protect the HEU appropriately or to conduct the research that once required the HEU. In addition to a number of facilities in Russia, these include facilities in Latvia, Belarus, Ukraine, Kazakhstan, and Uzbekistan, with quantities ranging from a few kilograms to tens or hundreds of kilograms of HEU. Proliferation risks could be rapidly reduced by purchasing these small, vulnerable stocks of HEU from these facilities and then blending them to low-enriched uranium (LEU) in Russia, either at the facilities performing blending for the larger HEU purchase or at other facilities that have such capabilities, such as the Luch Production Association at Podolsk. To facilitate rapid agreement on giving up such HEU stockpiles, assistance should also be provided for alternative research—including converting nuclear facilities that will still operate to use LEU rather than HEU.

This would require added funding of approximately $50 million.

Expand assistance for interdicting nuclear smuggling. The United States has provided assistance to install nuclear material detection equipment at three important transit points in Russia—Sheremetyovo airport in Moscow and two points on the Caspian Sea. Russia has requested equipment for 22 other key transit points. Providing this equipment and the necessary associated training would probably cost less than $20 million. More broadly, a strategic plan for interdicting nuclear smuggling efforts needs to be put in place, with appropriate funding, defining what groups in which countries should be provided what capabilities by when.

This would require added funding of approximately $20 million per year.

Stabilizing Nuclear Custodians

Target a portion of economic assistance to Russia to the nuclear cities. The United States continues to provide assistance to investment and economic reform in Russia because strengthening the Russian economy serves long-term U.S. interests. Such investments serve U.S. interests even more directly in cities where economic desperation could lead to the sale of nuclear knowledge or material. A portion of the economic reform assistance provided to Russia, from a broad range of U.S. programs, could be targeted to the nuclear cities, as is being done for other cities in the Regional Investment Initiative. This would complement but not replace the DOE's Nuclear Cities Initiative program, which should have a role in coordinating such efforts.

This would require added funding of approximately $30 million per year.

Monitoring Stockpiles and Reductions

FINANCE DATA EXCHANGES ON NUCLEAR MATERIAL STOCKPILES. Achieving a better understanding of the actual quantities, forms, and locations of fissile material in each country is fundamental to cooperative efforts to secure, monitor, and reduce these dangerous stockpiles (see the report of Task Force IV). The United States has openly published data on its plutonium stockpile and plutonium production and will soon publish similar data concerning its HEU stockpile. Providing the financing necessary for Russia to do the work of pulling together similar data, in return for sharing the data with the United States—starting with plutonium and, if that was successful, moving on to HEU—could offer a rapid means to accomplish part of the stockpile data exchanges agreed to by Presidents Bill Clinton and Boris Yeltsin in 1994 on a contracting basis, without requiring high-level formal negotiations.

This would require added funding of approximately $20 million.

FINANCE INTERNATIONAL MONITORING OF EXCESS FISSILE MATERIAL. The United States, Russia, and the International Atomic Energy Agency (IAEA) are pursuing a trilateral initiative designed to make possible international monitoring of U.S. and Russian excess fissile material without compromising proliferation-sensitive information. This initiative could play an important role in long-term controls over weapons-usable nuclear material and, by demonstrating to the world U.S. and Russian intentions that this material will never be returned to weapons, can help build political support for the nuclear nonproliferation regime. One key question is who will pay the costs to Russia and the IAEA of monitoring in Russia. (To date, the United States has been paying both its own costs and the IAEA's costs of monitoring the small amount of excess material that is under IAEA verification so far in the United States.) Agreeing to pay these costs could enable a significant nonproliferation and disarmament initiative to go forward, at a very modest cost.

This would require added funding of approximately $5 million per year.

Ending Production of Fissile Material

CONFIRM NONPRODUCTION OF HEU. In addition to fully funding the current effort to convert Russia's production reactors so that they no longer produce weapons plutonium, the United States could pursue agreement with Russia to establish reciprocal transparency measures at U.S. and Russian enrichment facilities to confirm that neither country is producing HEU. These measures could provide a test bed for approaches to verifying a fissile cutoff treaty.

This would require added funding of approximately $10 million per year.

Reducing Stockpiles of Excess Material

EXAMINE FEASIBILITY AND COST OF RAPID HEU BLEND-DOWN. Given the large size of the stockpiles of excess plutonium and HEU in Russia and the United States, most steps to address the risks they pose would cost more than the few millions or tens of millions included in the modest-steps category. One step that could be taken, however, is a detailed study outlining what would be needed to blend all of the excess Russian and U.S. HEU to non-weapons-usable levels within a few years, resolving rapidly the key nonproliferation and disarmament issues this material poses, even while it continues to be released onto the commercial market at a much lower pace. Blending to an intermediate level of perhaps 19 percent and postponing some of the purification until after that intermediate blending had been accomplished might allow more rapid completion of this initial blend-down. Significant capital investments in additional or modified blending capacity might well be required, as well as operational costs of implementing the rapid blend-down, which could be identified in such a study. Opportunities for using blended down HEU as collateral for loans or prepayments should also be considered in such a study.

This would require added funding of approximately $2 million.

Significant Strides

Preventing Theft and Smuggling

Expand MPC&A to a level not constrained by funding. There is a fundamental question that the U.S. government has not yet addressed in detail: what would it cost to create a situation in which lack of funds was not a significant constraint on the pace at which security and accounting for nuclear material in Russia could be improved? Achieving the most rapid and long-lasting practicable reductions in the proliferation threat posed by insecure nuclear material in the FSU would require providing adequate funding to (a) rapidly consolidate material in the smallest practicable number of buildings and sites; (b) provide both facility-level and national-level security and accounting system improvements as quickly as practicable; and (c) provide resources and incentives to sustain effective security over time (including ensuring that security and accounting upgrades are actually used and maintained).

The latter goal involves changing ways of thinking and patterns of organizational behavior, which is a challenge that involves much more than money. Efforts in that direction would range from paying for initial operations and maintenance of installed security and accounting systems, to strengthening regulators' ability to enforce security and accounting standards, to supporting broad MPC&A training and institutional reform programs, to providing assistance for regular and realistic performance testing of installed systems (and for fixing problems identified in such testing), to preferentially directing U.S. contracts to facilities with excellent security and accounting. Planning and activities for FY 1999 demonstrate that at least $150 million a year can be effectively spent on upgrades and training (the second of these tasks), and analyses within the MPC&A program and at the national laboratories suggest that fully effective programs in each of the other areas might require $50 million per year or more. Hence, $250 million per year for perhaps five years represents a *minimum* level for such a funding-unconstrained program; further creative thinking may identify opportunities that would require still larger levels of funding.

This would require added funding of approximately $100 million–$250 million per year.

Stabilizing Nuclear Custodians

EXPAND THE NUCLEAR CITIES INITIATIVE TO SUPPORT A BROAD RANGE OF SUSTAINABLE EMPLOYMENT. In thinking about the problem of providing alternative employment for excess Russian nuclear weapon experts, it is important to keep several distinctions in mind. There is a distinction between purely *commercial* employment (which may be difficult to achieve quickly, given Russia's economic crisis and the unique barriers to business development in the closed nuclear cities) and *civilian* employment (which could include, for example, nonweapons R&D sponsored by the Russian and U.S. governments and that may be easier to arrange in the near term). Another is the distinction between alternative employment for people who are leaving the weapons complex (where foreign assistance can play a substantial role) and sustaining those who will continue to maintain Russia's nuclear stockpile (which will remain Russia's responsibility). Yet another distinction is between the large number of employees in Russia's nuclear complex who play supporting roles and do not in themselves pose proliferation threats and the smaller number who have direct access to weapons-usable nuclear material or proliferation-sensitive information.

A comprehensive approach to providing alternative employment for excess nuclear weapons workers in Russia's nuclear cities would include, at a minimum, three elements. The first is a broad range of measures to support private-sector employment growth, ranging from business development centers to tax incentives for employment of excess nuclear weapons workers, including both establishment of new businesses in these cities and employment of nuclear-city experts as "knowledge workers" by foreign businesses. The second is support for employment of nuclear-city experts on nonproliferation and arms control analysis and technology development, thus providing employment well-matched to their nuclear skills while serving other U.S. arms control and proliferation interests as well. The third part of a comprehensive approach is support for employment of nuclear-city experts on tasks related to nuclear cleanup, energy, and the environment.

The second and third of these areas could be supported in a win–win approach by contracting a fraction of the hundreds of millions of dollars DOE spends each year on R&D in these areas to experts from the nuclear cities—thus getting the DOE's work done for less while providing interesting and relevant R&D employment to excess experts in the nuclear cities. In particular, work to develop safer, cheaper, and more proliferation-resistant approaches to nuclear energy and the fuel cycle—as MINATOM has proposed—would make use of the specific nuclear expertise of nuclear-city experts. In a report to Congress in early 1999, DOE estimated that the level of funding needed to create sustainable jobs for all the nuclear weapons and materials workers MINATOM expects to be displaced over the next several years would cost approximately $550 million over 5 years.[3] Yet, under the Expanded Threat Reduction Initiative, DOE envisions spending less than one-third that

---

3. *A Report to the Congress on the Nuclear Cities Initiative,* U.S. Department of Energy; reprinted with a range of other related documents and a summary of program status in *The Nuclear Cities Initiative: Status and Issues* (Washington, D.C.: Russian-American Nuclear Security Advisory Council, January 1999).

amount. Expanding this program enough to provide the needed employment could substantially reduce proliferation threats from unemployed nuclear weapons workers.

Not all of this money need necessarily go through DOE's Nuclear Cities Initiative program itself; it may well turn out that certain parts of the program can be carried out more effectively by institutions ranging from the International Science and Technology Centers (ISTC) to the European Bank for Reconstruction and Development (EBRD). At the same time, the United States should insist that Russia itself make substantial contributions to the effort to convert these cities to other tasks and place high priority on ensuring that remaining nuclear workers are paid.

The distinction between providing alternative employment for workers no longer needed for nuclear weapons work and relieving the desperation of those who still have access to nuclear materials and information will have to be carefully managed. A fundamental part of this effort would be the physical dismantlement or conversion of facilities for assembly of weapons and weapons components that are no longer needed.

This would require added funding of approximately $70 million–$100 million per year.

Monitoring Stockpiles and Reductions

Provide financial assistance for transparent warhead dismantlement. As described in the report of Task Force IV, increased transparency in the management of nuclear warhead and fissile material stockpiles (while maintaining protection for legitimate nuclear secrets) will be fundamental to achieving deep reductions in nuclear arms as well as cooperation to secure nuclear stockpiles—and hence to reducing the nuclear threat to the United States. Unfortunately, Russian secrecy concerns have largely blocked transparency progress in recent years, and increased concern over protecting nuclear secrets in the United States in the wake of the revelations of Chinese espionage is likely to make increased transparency a hard sell in the United States as well.

To date, the United States has not offered Russia any significant incentives—strategic, financial, or otherwise—to agree to accept wide-ranging transparency for warhead and fissile material stockpiles. The original Nunn–Lugar legislation offered the possibility of providing assistance for the actual dismantlement of nuclear warheads—which has not yet been done—if and only if there was transparency to confirm this dismantlement was taking place. Although Russia has never asked for financial assistance for warhead dismantlement (presumably in part to avoid being drawn into a dismantlement transparency discussion), Russian officials have repeatedly complained about the high cost to Russia of this dismantlement and have recently indicated that dismantlement rates have greatly declined, in part because of this high cost.

Under the Expanded Threat Reduction Initiative, the United States plans to offer some support for warhead dismantlement within a few years. The United States could offer to provide financial assistance for warhead dismantlement (e.g., $90 million per year for a dismantlement rate of 3,000 per year, or roughly $30,000 per warhead) in return for Russian agreement to a transparency package that would also be implemented reciprocally at the Pantex facility in the United States. If availability of adequate storage space for fissile material from so many dismantled weapons is a problem, additional assistance could be provided for more secure storage space—possibly making use of existing areas in guarded Russian nuclear facilities.

Both sides would insist that such transparency be implemented in a manner that did not unduly compromise warhead design information or sensitive data related to stewardship of the remaining nuclear weapons stockpiles. Such an offer of assistance for transparent dismantlement would have a better chance of being accepted as part of a nuclear reductions package that addressed some Russian concerns (such as concerns over the U.S. upload "hedge" stockpile) perhaps in return for Russia addressing some U.S. concerns (such as the Russian tactical nuclear stockpile).

This would require added funding of approximately $90 million per year.

Ending Production of Fissile Material

Finance verification of a fissile cutoff in Russia. If successful, the fissile cutoff agreement now being discussed at the conference on disarmament in Geneva would be a significant accomplishment, ending forever mankind's production of fissile materials for weapons. It would require, at a minimum, IAEA verification at all reprocessing and enrichment plants to confirm that they are not producing material for weapons. Verification at older reprocessing plants never designed for safeguards, where inspectors cannot gain access to large, remotely handled areas of the interior of the plant because of the intense radioactivity, is likely to be particularly problematic. The costs of verification in Russia, both to the IAEA for its inspections and to Russia to prepare for and host these inspections, are likely to be significant. Russian inability or unwillingness to pay these costs is likely to interfere substantially with successful negotiation of such an agreement (and, indeed, may be part of the motivation for current Russian proposals that would exempt enough material from verification to render such an agreement virtually meaningless).

The United States could offer to finance the costs of this verification—directly or, for example, by contributing to an IAEA fund that would finance these costs. The cost would likely be in the range of a few tens of millions of dollars per year, perhaps with somewhat larger start-up investments. Such an offer would be a sensible investment in security, significantly increasing the chances of success in negotiating a fissile cutoff and laying the basis for improved accounting and control of weapons-usable material at Russian reprocessing and enrichment facilities, that will play a fundamental long-term role in both ensuring adequate security and being part of a broad regime for verifying deep reductions in fissile material stockpiles. It would also make sense to begin immediately to carry out small-scale cooperative experiments at U.S. and Russian reprocessing plants to demonstrate technologies and procedures for verifying a cutoff at older reprocessing plants.

This would require added funding of approximately $30 million per year.

Reducing Stockpiles of Excess Material

Finance rapid blend-down of excess HEU. The current 30-ton-per-year pace of blending in the U.S.-Russian HEU purchase agreement is determined by the rate at which the commercial market can absorb the material, not the rate that would best serve U.S. security interests. From a security perspective, it would be desirable to blend all excess HEU to non-weapons-usable form immediately. As noted above, such rapid blending—perhaps to an intermediate level of 19 percent, leaving the final blending to 4 percent and chemical purification to be done at the current market-driven pace—would require some up-front capital investment in blending facilities. There would be operations costs as well, and some additional financial incentives might be required to gain Russian agreement to carry out this operation. The material would continue to be released on the market at a pace consistent with commitments made to uranium industry firms as part of the existing HEU purchase agreement. All told, the cost of achieving the blend-down of all excess Russian HEU within, say, five years might come to something in the range of half a billion dollars—a small price for the large security benefit of accomplishing this objective.

This would require added funding of approximately $100 million per year.

Finance construction of necessary facilities for plutonium disposition. The task force concurs with numerous previous reports—including the 1997 special CSIS panel report on this issue—that reducing excess plutonium stockpiles as rapidly as practicable is a high priority. Plutonium disposition is in the unusual position of being a long-term issue requiring urgent action—in part because the U.S. Congress has made clear that it will not support planned requests for funding for construction of plutonium disposition facilities in the United States in the absence of substantial progress toward a clear commitment that Russian plutonium stockpiles will be reduced in parallel. Although considerable progress is being made in U.S.-Russian discussions regarding such commitments, the fundamental issue of who will pay the more than $1 billion cost of disposition of the Russian plutonium—including the cost of needed plutonium fuel fabrication facilities and reactor modifications for using plutonium fuel—remains open. This amount is small in comparison with the potential threats to international security posed by this excess plutonium; and if an agreed plan can be worked out that would make it possible to eliminate this excess plutonium stockpile on a reasonable timetable, it would make sense to fund this investment over a period of, perhaps, five years.

The first priority should be to ensure that all the excess plutonium is secure, placed under monitoring, and converted to unclassified forms as rapidly as practicable, pending longer-term disposition. The U.S. Department of Defense is already negotiating with Russia to arrange support for the conversion of plutonium weapons components to unclassified metal forms, and the DOE is planning to provide support for converting those metal forms to oxides that would be suitable input for the various longer-term disposition options.

This would require added funding of approximately $200 million per year.

Great Leaps

Preventing Theft and Smuggling

PROVIDE INCENTIVES TO CONSOLIDATE HEU AND PLUTONIUM IN INTERNATIONALLY GUARDED FACILITIES. International confidence in the security of nuclear material in the FSU (or anywhere else for that matter) would be greatly increased if the material were in a storage facility with *international* guards and monitors instead of solely national ones. One possibility would be to establish one or more internationally guarded storage facilities in Russia and create a fund that would pay Russia a substantial sum—for example, $10,000 per kilogram, roughly half the price for buying blended HEU in the U.S.-Russian HEU purchase agreement—to deposit fissile material at this facility. The actual physical facility would presumably be the Mayak storage facility now under construction with U.S. assistance, possibly with the construction of additional modules to accommodate all the material that might be involved in such an initiative.

The advantage of offering the incentive for storing material in the facility, compared with simply building the Mayak storage facility and leaving it at that, would be the agreement to have an international guard force and international control over the facility. Parallel facilities with similar international guard arrangements could be established in the United States for reciprocity (likely to be essential if there is to be any hope of Russian agreement to such an approach) and, ultimately, in other countries with large fissile material stockpiles as well.

In an extreme case, if Russia deposited 700 tons of HEU and 100 tons of plutonium in such a facility, the cost of the financial incentive would be $8 billion. Alternatively, the direct payment for deposit could be limited to plutonium (reducing the cost of the incentive in the previous case to only $1 billion), with the storage facility serving as a way station for HEU being stored there prior to blending for sale. There would be continuing annual costs in the range of tens of millions for the international guarding and monitoring of the site or sites. There would inevitably still be other facilities where plutonium or HEU was present, but this approach would potentially offer the opportunity to provide high-confidence security and transparency for a large fraction of the FSU fissile material stockpile within a few years.

This would require added funding of approximately $1 billion for excess plutonium, increasing up to an additional $7 billion (approximately) if the same incentive was offered for all excess HEU.

Stabilizing Nuclear Custodians

ESTABLISH A COMPREHENSIVE PROGRAM TO DOWNSIZE THE RUSSIAN NUCLEAR COMPLEX AND PROVIDE ALTERNATIVE EMPLOYMENT. A truly comprehensive program would involve agreeing on a strategic plan for downsizing the Russian nuclear complex—both the closed cities of the weapons complex and the widely scattered civilian facilities—to a size that was sustainable and appropriate for the post–Cold War world, physical dismantlement or conversion of the facilities no longer needed under such a plan, and provision of sustainable employment opportunities to workers who might otherwise be tempted to sell their nuclear knowledge or nuclear materials to which they have access. For such a plan to be successful, it would have to be developed jointly with senior Russian nuclear officials, who would have to take the lead in its implementation. The cost of such a comprehensive approach has not yet been estimated but would likely be significantly higher than the $550 million estimated by the DOE for a nuclear-cities-only downsizing and economic stabilization program.

This would require added funding of approximately $1 billion–$5 billion.

Monitoring Stockpiles and Reductions

PROPOSE AND FINANCE A WIDE-RANGING RECIPROCAL WARHEAD- AND MATERIALS-MONITORING REGIME. Over the long term, U.S., Russian, and international security would be well served by building step by step toward a comprehensive transparency regime for warheads and fissile materials. Such a regime would ultimately include the presence of monitors or electronic remote monitoring systems for all fissile material, providing real-time transparency for the entire stockpile (see the report of Task Force IV) as well as real-time confirmation that material had not been removed without authorization. Agreement on such a regime is difficult to imagine in today's political atmosphere between the United States and Russia but might become possible in the longer term, after other cooperative steps had proved successful. Transparency measures that would contribute to such a regime can be put in place step by step, with each step offering some benefit while posing modest risk. Although the full cost of such a comprehensive regime has not been assessed in detail, it is likely to be significant, particularly if both the United States and Russia retained nuclear complexes similar in scale to today's.

This would require added funding of approximately $100 million–$500 million per year.

Ending Production of Fissile Material

None of the measures that appear necessary to address the goal of ending production of fissile material would be likely to require funding in the range of billions of dollars.

Reducing Stockpiles of Excess Material

PURCHASE ADDITIONAL QUANTITIES OF EXCESS HEU. Russia has far more HEU that will ultimately be excess to its military needs than the 500 tons the United States has agreed to purchase in the U.S.-Russian HEU purchase agreement. The United States or other countries could offer to buy additional stockpiles of excess HEU, with government funds, to be held off the market as a strategic uranium stockpile (rather than flooding the market immediately with still more material from excess HEU available for direct market sales). As a first step, the United States or another government could offer to buy an additional 100 tons of HEU—either blended, if that can be done quickly, or shipped as HEU for blending in the United States, if that would be faster and Russia would be willing. If that seemed to work well, an additional 100 tons could be purchased.

Ultimately, it would make sense to buy every bit of excess HEU Russia was willing to make available. At the prices the United States and Russia originally negotiated for the HEU purchase agreement ($12 billion for 500 tons), another 100 tons would cost $2.4 billion. (It would be worth significantly less in unblended form because a significant fraction of the purchase price is actually the value of the blendstock.) It might be possible to specify in the contract that some portion of the proceeds be used for nuclear security activities (see discussion in appendix D of this chapter, "Generating New Revenue for Nuclear Security," page 23).

This would require added funding of approximately $2.4 billion–$12 billion.

PURCHASE RUSSIAN EXCESS PLUTONIUM. On today's commercial nuclear fuel market, plutonium has no value because making fuel even from free plutonium is more expensive than buying equivalent uranium fuel on the open market. But Russia considers its excess plutonium a major national asset produced with thousands of work-years of socialist labor—and from the point of view of investing in security, it makes every bit as much sense to invest in securing plutonium as it does to invest in securing HEU. If the idea of a financial incentive to deposit the plutonium in an internationally guarded facility does not pan out, simply offering to buy Russia's excess plutonium outright should be seriously considered. Even if plutonium is overgenerously valued at the same price as the equivalent fissile quantity of HEU, buying the entire 50-ton stockpile of plutonium Russia has so far indicated is excess to its military needs would cost $1.2 billion, while buying the 100-ton stockpile that might become excess as additional reductions are agreed to would cost $2.4 billion.

The political difficulties (in both Russia and the United States) of actually shipping the material to the United States after the purchase might make it simpler to keep the material in Russia and place it in a storage facility with a guard force on the U.S. payroll pending its eventual disposition—which could be by any method the new owner of the material determined was most desirable. The new owner of the material would presumably also have to pay for its storage and disposition—unless the international community could be convinced to chip in—which could add up to several billion dollars over the long run.

This would require added funding of approximately $1.2 billion–$5 billion.

Leadership and Synergies

A wide range of cooperative activities is already under way to secure, monitor, and reduce nuclear stockpiles in the FSU and the United States. A long menu of additional possibilities is available if additional resources are provided.

To make all of these efforts work as a package; to coordinate, prioritize, and integrate them into a strategic plan; and to negotiate them with Russia while overcoming the many obstacles to expanded cooperation would require a dramatic increase in sustained leadership from the highest levels of the U.S. government. In particular, it will require designating a full-time point person for this task, as was recently done for overseeing implementation of the Agreed Framework with North Korea. Preventing nuclear material from falling into the hands of states like North

Korea or Iraq is certainly no less critical to U.S. security; indeed, the entire global effort to prevent the spread of nuclear weapons depends on it. The cost of taking action now to address this threat is tiny in comparison with the cost and risk of failing to act and finding that the essential ingredients of nuclear weapons find their way into the hands of terrorists or proliferant states.

Appendixes

A. Current Programs

Dozens of cooperative programs with the states of the FSU (particularly Russia), costing hundreds of millions of dollars per year, are already under way to address the security risks that are the focus of this task force report. These programs represent a critical investment in U.S. national security and deserve strong support. Many of these efforts are making tremendous progress—but wide gaps still remain between what has been accomplished and what needs to be done to address the threat. Although an excellent start has been made in many areas, most of the work remains to be done.

The MPC&A program—designed to improve security and accounting for all weapons-usable material in the FSU—is now engaged at virtually every site in the FSU where such material is known to exist, with a budget in the range of $140 million a year. Security for tens of tons of material at more than a dozen sites has been demonstrably improved, but there are dozens of sites with hundreds of tons of material remaining. The issue of how to ensure that security that is sustainable for the long term can be achieved remains a fundamental question. A substantial program is under way to cooperate in improving security and accounting for nuclear warheads as well, during both transport and storage, but here, too, most of the needed upgrades remain to be accomplished. With U.S. assistance under DOD's Cooperative Threat Reduction program, known as the Nunn–Lugar program, a modern, secure facility for storing plutonium and HEU from dismantled weapons is under construction at the Russian facility known as Mayak and should be open by 2002. Transparency measures are not yet agreed, and more storage capacity would be needed to store all the excess material from Russian dismantled nuclear weapons. (A new umbrella agreement extending the Nunn–Lugar program years into the future was signed in mid-June 1999, providing an excellent basis for continued progress in many of these programs.)

Several U.S. and international programs provide funding for alternative employment for nuclear weapons scientists and engineers. The International Science and Technology Centers are employing some 24,000 former experts on weapons of mass destruction at least part-time on short-term research grants. The DOE's Initiatives for Proliferation Prevention program provides similar grants for projects that seek to match Russian technologists to U.S. laboratories and U.S. industry, but its success in commercializing technologies has been very modest to date. The new Nuclear Cities Initiative is intended to provide a broad approach to developing alternative employment in the nuclear cities, but it faces daunting challenges and the funding provided to date is modest. The largest flows of money to the desperate Russian nuclear complex from U.S. programs do not come from any of these efforts but come instead from the U.S.-Russian HEU purchase agreement (described below), which will earn Russia more than $500 million per year.

To build transparency in the management of nuclear weapons and materials, the United States and Russia have agreed in principle to a wide variety of exchanges of data and reciprocal monitoring, but few of these have been implemented—largely because of Russian concerns over providing sensitive information and access to sensitive facilities. The only warhead or fissile material transparency measures currently being implemented on a large scale are those for the HEU purchase agreement, designed to build confidence that the LEU the United States is purchasing comes from HEU from dismantled weapons and that the United States uses the purchased material only for peaceful purposes. Promising laboratory-to-laboratory work is under way, however, to develop joint approaches to confirming warhead dismantlement and monitoring warhead stockpiles while protecting sensitive information in the hope of providing confirmation and monitoring tools that will be available to negotiators when formal negotiations begin.

A substantial U.S.-Russian effort is under way to convert the cores of the remaining Russian plutonium production reactors—which continue to operate and produce plutonium because they also produce essential heat and power for nearby communities—to a new fuel that will not require reprocessing and will produce far smaller quantities of poorer-quality plutonium. If all the relevant safety, security, and licensing issues can be resolved in time, these reactors are to be converted during 2000–2002, ending Russia's production of separated, weapons-grade plutonium.

Separate programs are under way to reduce stockpiles of excess HEU and of excess plutonium. HEU is to be blended with other uranium to produce LEU for sale on commercial nuclear fuel markets. In the U.S.-Russian HEU purchase agreement, the United States is buying 500 tons of Russian HEU from dismantled weapons—blended to LEU over 20 years—for a price originally set in the range of $12 billion and then reselling the material on the commercial nuclear fuel market. This is a remarkable initiative that provides a financial incentive for warhead dismantlement, destroys enough weapons material for many thousands of nuclear weapons, and creates a large revenue stream for the desperate Russian nuclear complex, all at little net cost to U.S. taxpayers. Russia will ultimately have far more than 500 tons of HEU that is no longer needed for its military programs, however.

Unfortunately, problems that arose from the privatization of the main U.S. implementer of this agreement—the U.S. Enrichment Corporation—placed this major national security initiative at the mercy of the commercial marketplace; therefore, it did become necessary to provide a one-time government subsidy of $325 million in FY 1999 to keep the deal moving forward. That subsidy led to a new agreement that reinvigorated implementation of the agreement.

The task force strongly urges the U.S. government to place very high priority on ensuring that this deal continues to move forward in the future—an issue that is likely to arise again soon, as the current contract for purchases of the enrichment component comes up for renegotiation. At the same time, the United States is beginning to blend down 174 tons of its own HEU, which it has declared excess.

For plutonium, the situation is more complicated. Unlike HEU, plutonium is a liability in the current nuclear fuel market because handling plutonium is so expensive that fuel made from plutonium is more expensive than uranium fuel bought on the open market, even if the plutonium itself is free. Moreover, nearly all isotopes of plutonium are weapons usable, so the proliferation risks posed by plutonium cannot be easily removed by blending as they can with uranium. The U.S. government has decided to take two approaches in parallel for reducing its own excess plutonium stockpiles, paying the extra price to use some of the material as fuel in nuclear reactors and immobilizing the remainder with nuclear wastes. Both approaches result in the plutonium being embedded in massive, highly radioactive waste forms that would be difficult to steal and from which it would be difficult and costly to recover the plutonium, and both approaches represent only modest additions to the stockpiles of nuclear waste that must be disposed of in any case.

Russia has rejected the immobilization option, preferring to use its excess plutonium as reactor fuel. Russia has no funds available to support the expense of disposition of excess plutonium, and the number of safe, modern reactors available in Russia may not be enough to burn the excess plutonium at a reasonable rate. U.S.-Russian negotiations on a plutonium disposition agreement are under way, but questions of how disposition will be financed and what reactors would be used still remain open.

In January 1999, President Clinton announced a new Expanded Threat Reduction Initiative designed to expand efforts to work cooperatively with Russia on reducing these and other dangerous leftovers of the Cold War. In the case of programs related to warhead and fissile material controls, the initiative called primarily for continuing funding for programs that had once been expected to be nearly complete by now—because increasing cooperation has revealed major new areas of work that needs to be done—rather than making major increases from previously appropriated levels. For example, for MPC&A—the program most directly related to safeguarding weapons-usable nuclear material—Congress appropriated $152 million in FY 1999 and, after internal DOE reallocations, the program actually received $140 million; the request for FY 2000 under the new initiative is $145 million. Thus, while the Expanded Threat Reduction Initiative is a worthy effort deserving support, it alone will not address the gap between the scale of the current effort and the scale of the U.S. security interests involved.

B. Sizing the Problem

While the problems posed by Russia's nuclear infrastructure are enormous, they are not infinite. A few back-of-the-envelope calculations provide useful bounds for thinking about the magnitude of the problem.

- **Costs of safeguards and security.** The weapons-usable nuclear materials in the FSU are believed to exist in more than 300 buildings at nearly 60 sites. If there was no progress at all in consolidating this material at fewer sites, and it cost $10 million, on average, to upgrade security and accounting for each of these buildings, the total cost of these upgrades would be in the range of $3 billion. The costs of guard forces and operations, maintenance, testing, and improvement of these security and accounting systems would probably run to hundreds of millions of dollars per year. These are large numbers—demonstrating the need for consolidation to reduce costs—but not outlandish ones. Russia, for example, receives more than $500 million each year from income on the HEU deal alone. The United States spends more than $700 million a year on the broad spectrum of safeguards and security activities (not just guarding fissile material) in the DOE complex.

- **Costs for paying nuclear weapons complex employees.** Russia's vast nuclear weapons complex still includes 10 entire closed cities—cities built entirely for the purpose of designing and producing nuclear weapons and their ingredients—that remain fenced off from the outside world and surrounded by armed troops. There are a total of approximately 125,000 employees of the nuclear facilities in these 10 closed cities. Currently, on average, they are reportedly paid less than $100 per month. The total payroll for the nuclear facilities of the closed cities is therefore in the range of $150 million a year—again, far less than what Russia takes in each year from the HEU deal alone (although not all of those funds go to MINATOM).

  The average age of these workers is now between 50 and 60, and, with the precipitous decline of male life expectancy in Russia, this workforce is likely to decline fairly rapidly through attrition alone in the years to come. The total cost for paying for all of the Russian nuclear weapons complex workers to retire on full salary for the rest of their lives would probably be in the range of $2–$4 billion. Of course, some of these workers could be paid to work on decommissioning and cleaning up the sites where they used to work on producing weapons and materials rather than be paid to retire. Such a combination of buyouts to encourage early retirement and engaging workers in cleanup is, in effect, the approach that has been taken in shrinking the nuclear weapons complex in the United States—but Russia has not had the funds to pursue a similar strategy.

- **Commercial value of Russia's total stockpiles of fissile material.** By unclassified estimates, Russia is believed to have roughly 1,050 tons of HEU and 160 tons of separated plutonium—in round numbers, approximately 1,200 tons of fissile material. At the price—$12 billion for 500 tons—originally negotiated for the HEU purchase agreement, the value of the entire stockpile, including all the material in all of Russia's nuclear weapons, would be in the range of $29 billion—less than the costs of many of the new weapon systems currently planned by the U.S. Department of Defense.

C. Obstacles to Nuclear Security Cooperation

There are enormous obstacles to genuine cooperation in sensitive nuclear security areas between former adversaries like the United States and Russia, which remain deeply suspicious of each other's motives. The souring of U.S.-Russian political relations during 1998 and 1999, accelerated by the North Atlantic Treaty Organization's bombing of Yugoslavia, has made cooperation even more difficult and the hurdles to be overcome if major new steps are to succeed even higher. The experience of the past several years makes clear that nuclear security cooperation can succeed only if it is approached as a genuine partnership, with experts from both sides contributing their work and ideas to solve common problems together—rather than as an effort by one side to impose solutions on the other.

The most fundamental obstacle to cooperation is that the United States and Russia continue to have many conflicting interests—although they have profound common interests in preventing the proliferation of nuclear weapons and achieving permanent nuclear arms reductions. Russia would like to maintain a cutting-edge nuclear arsenal comparable with that of the United States, and the United States has no interest in helping Russia do that. Both countries' intelligence services would like to find out as much as possible about the other country's nuclear secrets, and both countries are deeply suspicious of the other's intentions in that regard. The United States has an interest in achieving specific nonproliferation and arms reduction goals as quickly and cost-effectively as possible. Russia, while not opposed to that objective, also has an interest in providing as much employment for excess nuclear workers as possible for as long as possible. The United States has an interest in seeing the assistance it provides also support U.S. contractors and laboratories. Russia has an interest in seeing as much of that money as possible go to Russian entities. Substantial segments of the political establishments in both countries—including a large fraction of both countries' legislatures—remain deeply suspicious of the other and skeptical of the whole idea of nuclear security cooperation (which creates a constant danger that problems that arise will be blown out of proportion and spin out of control). Given these differences of interest, disagreements about specific approaches to cooperation are inevitable and need to be resolved patiently with good-faith negotiation and discussion.

Secrecy and limited access to facilities are serious obstacles faced by essentially all cooperative nuclear security programs with Russia. Enormous progress has been made in breaking down barriers, particularly compared with the early 1990s when Russia would not allow cooperation on security upgrades at *any* site, even civilian ones, with separated plutonium or HEU. Today cooperation is under way at almost *every* such site. Enormous barriers still remain, and, as political relations have deteriorated, the Russian security services have become more active in restricting access, even to facilities where access had been granted before. At the same time, the United States has frequently been reluctant to offer comparable access to its own facilities, and this, too, has slowed progress. The recent furor over Chinese spying and laboratory security in the United States will only make this problem worse; if approved, some of the draconian restrictions on the U.S. laboratories' foreign interactions that have been proposed could effectively destroy any prospect for effective cooperation to address these security threats.

Competing priorities, bureaucratic disorganization, frequent changes of government personnel, and lack of sustained attention to these issues by the highest levels of government have been serious problems on both sides. It is difficult to do business with a Russian government facing a thousand priorities it considers more urgent, whose prime minister changes every few months, whose ministries often do not communicate, whose nuclear facilities increasingly may not abide by deals cut in Moscow, and whose senior leadership takes only occasional interest in resolving critical issues related to these nuclear security cooperation programs. Russian experts have much the same complaints about dealing with the U.S. government.

Another major issue is the difficulty of ensuring that U.S. taxpayers' dollars are being spent as they should be—an issue with several parts. First, there is the widespread corruption in Russia, which makes it essential to structure assistance programs so that the funds cannot simply be raked off into foreign bank accounts. (Contrary to popular impressions, U.S. programs in these areas virtually never simply fork over large quantities of cash in return for a promise that it will be spent in particular ways; rather, funds go to procure particular equipment needed for a job, or contracts are made for specific, demonstrable deliverables, such as building a fence in a particular spot.)

Second, Russia has a dysfunctional payments system in which, for example, money deposited at a particular bank for use by a joint project at a nearby institute may be seized by the bank to cover the institute's bad debts, or may be seized by tax police to cover the institute's back taxes, or may be used by a desperate institute director to pay salaries of other employees (in the hopes that it can be paid back if the institute's promised government funding ever comes).

Third, there is the problem of ensuring that equipment provided for a particular purpose is in fact used for that purpose and does not go unused (because there is no money to operate and maintain it) or get diverted to other purposes (which the recipients may consider higher priority). Fourth, there is the nettlesome—and growing—problem of Russian efforts to impose a variety of taxes, tariffs, and duties on U.S. assistance, in effect directing a portion of the assistance away from the agreed projects and into the general coffers of the Russian government instead. Although the specific situation of each nuclear security program is unique, all of them have faced these problems, and even when they have found successful solutions, the fact is that an enormous amount of time, energy, creativity, and political capital has had to be spent on procedures to ensure accountability rather than on getting the cooperative work done.

Cultural differences and poor negotiating tactics on both sides have, on occasion, also led to obstacles and disputes, which in some cases have delayed progress by months or years. What may appear from a U.S. perspective to be the minimum necessary audit and examination approach to ensure that U.S.-financed equipment is used appropriately may appear from a Russian perspective as unwarranted intrusion and possibly an intelligence mission. A policy change seen on the U.S. side as tightening up lax spending practices of the past may be seen on the Russian side as abrogating the spirit of partnership by ignoring Russian suggestions for how funds should be spent. Failure to include experts who can seriously address the technical aspects of these issues on negotiating teams can often present a problem, as can failure to take Russian experts' views fully into account.

Given this list—which is by no means comprehensive—of obstacles to cooperation, success in nuclear security cooperation is never guaranteed, and the obstacles to initiating major new efforts are substantial. The fundamental ingredients of success are sustained and energetic leadership; a genuine commitment to working in partnership; a step-by-step approach designed to build trust as progress is made; patience, persistence, and creativity in overcoming obstacles; and consistent follow-through on commitments. With those ingredients, and with a willingness to apply additional financial resources, there are opportunities for dramatic new progress in dealing with the nonproliferation and arms reduction challenges both countries face and perhaps even contributions to improving the overall political atmosphere between Russia and the United States.

D. Generating New Revenue for Nuclear Security

Essentially all of the funding for nuclear security in the FSU has come from government budgets—either the governments of the former Soviet states or foreign governments providing assistance. This is likely to continue to be the dominant source of funding for these activities in the future as well. But there may be opportunities to provide additional sources of revenue that would help improve security in the near term and help the former Soviet states maintain security as foreign assistance phases down over the longer term. At the same time, it is essential to continue to emphasize the importance of Russia itself contributing to these programs—providing support in kind (labor, use of facilities, HEU, plutonium) where on-budget funding is not available.

- **Spent fuel storage.** A variety of approaches have been proposed in which Russia would establish an international storage facility for spent nuclear fuel from a variety of countries and some portion of the profit would be set aside for nonproliferation and disarmament purposes, ranging from secure storage and monitoring of nuclear material to disposition of excess plutonium. (This differs substantially from MINATOM's proposed approach, in which the profit would be set aside to build and operate a large reprocessing plant.) One of these schemes has reached the point of detailed discussions of actual contracts and could potentially provide hundreds of millions of dollars per year for nonproliferation activities. The income from spent fuel storage should be sufficient to ensure that the storage would be safe and effectively safeguarded. (See the report of Task Force II).

- **Additional HEU sales.** As suggested in the text, the United States should seek to buy additional quantities of excess Russian HEU, above and beyond the 500 tons it is currently purchasing. As part of such an additional purchase, the United States should seek to require in the contract that a substantial fraction of the proceeds be spent on specified nuclear security purposes—ensuring nuclear guards and workers are paid, operating and maintaining security and accounting systems, and the like. (If the idea is presented carefully, MINATOM may be favorably disposed to agree to such a requirement because it would help MINATOM ensure that the funds stay within MINATOM rather than go to the rest of the Russian government.) Although confirming that the funds were spent as agreed would be an issue, there are past precedents of U.S.-Russian agreements with similar requirements. If Russia agreed to spend half the proceeds from the purchase of an additional 100 tons of HEU on nuclear security, this would make available more than $1 billion for these purposes.

- **A DEBT-FOR-SECURITY SWAP.** Russia is heavily burdened with foreign debt. Some restructuring of that debt is likely to be essential to economic recovery. In many less developed countries, foreign governments or organizations have negotiated debt-for-nature or debt-for-environment swaps, in which either a specified area of land is set aside as protected area or a certain quantity of money is set aside in a fund for environmental purposes in return for forgiveness of a certain quantity of debt.

  Some of these have already been successful in the former Communist states. For example, in 1991 the 17 creditor nations of the Club of Paris agreed to forgive 10 percent of Poland's debt if Poland instead set aside an agreed amount of money (less than the amount of debt forgiven) for environmental protection. Poland established an independent foundation, the Ecofund, to administer the funds, so that the amount of money in fact being spent could be easily verified.

  A similar approach could be taken for nuclear security, with a certain portion of Russian debt being forgiven in return for Russia agreeing to set aside funds for nuclear security in a similar independent fund. If Western governments are willing to forgive a substantial quantity of Russian debt, this could potentially provide a large enough revenue stream to support the hundreds of millions of dollars per year that it will ultimately cost Russia to ensure high levels of security and account for all of its nuclear weapons and fissile material.

- **DISCOUNT FROM AN INTERNATIONAL REPOSITORY.** A variety of groups are working to establish international storage sites or permanent geologic repositories, which would accept spent fuel from a variety of countries on a commercial basis. In at least one case (the group, known as Pangea, seeking to establish an international geologic repository in Australia), there has been a suggestion that in order to strengthen the nonproliferation and disarmament case for building such a repository and thereby further its chances of political approval, the Australian group would be willing to provide a substantial discount for disposal of spent fuel that originated as HEU or plutonium from weapons—with the idea that this discount, rather than going primarily toward a lower price for the utilities using this fuel, would largely go into a fund that could be used to finance nonproliferation and disarmament needs in the FSU. With a discount of $500 per kilogram of spent fuel, at the current U.S.-Russian HEU purchase agreement rate of 30 tons of HEU per year, more than $250 million per year might be available from such a discount; if the pace of the HEU deal could be increased to

50 tons of HEU per year, as much as $450 million per year might be available. This would be sufficient, for example, to pay for construction of facilities for disposition of excess weapons plutonium within a few years.

- **Encouraging MINATOM exports within proliferation constraints.** Much of the revenue of MINATOM now comes from exports of nuclear material, services, and technology, with MINATOM officials estimating that their total exports now amount to more than $2 billion per year. (The division of this revenue between MINATOM and the Russian central government is not well understood.) MINATOM is carrying out a variety of nuclear reactor exports opposed by the United States on proliferation or safety grounds (including reactors to Iran, India, and Cuba, among others), largely because of its need for export income. MINATOM's exports to less proliferation-sensitive markets are constrained by trade restrictions on Russia's low-cost uranium and enrichment exports imposed in the United States, Europe, and some other major nuclear markets and, in the case of reactors, by the widespread desire for Western nuclear reactor technology, rather than Russian technology, among those countries with access to it.

  Western countries could take a number of steps to reduce these export hurdles in return for specific commitments from MINATOM to apply particular fractions of the resulting revenue to identified nuclear security endeavors: easing trade restrictions on their uranium and enrichment markets; encouraging an expansion of Russian fuel fabrication service exports; and, ultimately, joint design of a new generation of reactors (such as the current Russian-Japanese-French-U.S. cooperation to develop a new high-temperature gas-cooled reactor). All of these would be sensitive because every piece of market share Russia gains would presumably be lost by a Western supplier. The dream of joint design of a new generation of reactors that could be exported to countries all over the world—often described at MINATOM headquarters—is not likely to come to fruition. The markets for new nuclear reactors are likely to remain limited; full development and demonstration of such new systems in Russia are likely to be more expensive than governments will be willing to support in the near term; and the continuing controversies over proliferation-sensitive Russian exports are likely to limit the possibilities of cooperation as long as those exports continue.

CHAPTER 2

An International Spent Fuel Facility and the Russian Nuclear Complex

Task Force II

THE INTERNATIONAL NUCLEAR AND ARMS CONTROL COMMUNITIES ARE FACING TWO COMPETING CHALLENGES THAT MERIT HIGH-LEVEL ATTENTION.

First, significant challenges need to be overcome to help ensure that sensitive nuclear materials and facilities in Russia are placed under far better control. This is a subject that urgently requires much higher sustained attention in the United States and elsewhere. There is an urgent need to identify credible resources that can be made available to enable Russia to move forward more aggressively with a comprehensive program to protect better and account for existing stocks of sensitive materials and to convert its sizable stocks of excess weapons plutonium to the spent fuel standard.

Second, within the civil nuclear sector, there is a need to identify and develop additional options that will enable states with rapidly accumulating inventories of spent fuel to manage better these inventories under conditions that will be fully compatible with nonproliferation and environmental values.

Fortunately, some new ideas are being proposed that hold out some hope of constructively contributing to both of these objectives. For example, suggestions have been made that serious attention should now be given to the merits of establishing some international facilities that would store spent fuel from several countries and possibly also dispose permanently of nuclear wastes. Proposals are surfacing that international spent fuel and/or waste facilities should be established in a host country with the idea that a substantial fraction of the profits might be devoted to improving the security of nuclear materials, including excess weapons plutonium in Russia.

It is a conclusion and recommendation of our CSIS working group that the U.S. executive branch and the Congress should give encouragement to the development and elaboration of these concepts, recognizing that, under U.S. law, U.S. prior approvals will be required before any U.S.-origin special nuclear materials can be transferred from any countries to hosting states.

It is recognized that if any international spent fuel storage facilities or waste repositories are to emerge from the current proposals they will have to satisfy several standards and criteria. The locations involved will have to be technically or geologically suitable for storing or disposing of spent fuel as nuclear wastes. The host organizations involved will have to have in place the requisite physical, technical, legal, administrative, and regulatory infrastructure to accommodate safely the protracted storage of spent fuel or the permanent disposal of nuclear wastes. In addition, the host countries will have to possess solid nonproliferation credentials, be politically acceptable to all interested states involved, and be perceived as stable and credible long-term custodians for the receipt of spent fuel or nuclear wastes.

There will have to be high confidence that the interested parties involved will rapidly be able to come to mutually agreed understandings as to the conditions pursuant to which the spent fuel and/or nuclear wastes would be transferred to the host country, including nonproliferation conditions. In the case of spent nuclear fuel, supplier states like the United States will have to be willing to consent to the transfer of the spent nuclear fuel to the host country. This will require a meeting of the minds as to the safeguards and other nonproliferation conditions that will apply to the materials involved.

Finally, there will have to be high confidence that the arrangements for spent fuel storage or nuclear waste disposal will be well understood and acceptable to the public of the host country and to neighboring states.

Satisfying these conditions will require concerted high-level efforts by the governments involved.

The Issues

Task Force II addressed the following major questions:

- Should the United States and other parties support Russian security efforts to protect better nuclear materials and expertise through the establishment of international or regional facilities for storing or disposing of spent nuclear fuel and/or nuclear wastes?
- Can such facilities produce revenues significant enough to reduce the nuclear dangers posed by the breaking up of the Soviet Union (through efforts to convert stocks of excess weapons plutonium to more proliferation-resistant forms)?

Goals

There is an urgent need for the United States to intensify its cooperative efforts with the Russian Federation to help Russia minimize the risk that nuclear materials or expertise will be diverted to military use, while both countries secure stable, economical, and clean energy supplies for the twenty-first century.

One of the gravest threats to the security interests of the international community, including the United States, stems from the inadequate controls that are being applied to sensitive nuclear materials and nuclear facilities, including excess nuclear weapons materials, in the Russian Federation. Although Russia is working with the United States and several other countries to address this critically urgent problem, far greater priority, substantially more resources, and far more vigorous political leadership need to be devoted to this subject.

For example, continued and assured progress in Russian and U.S. efforts to convert many tons of weapons plutonium to more proliferation-resistant forms will depend on the prompt identification of new sources of financing to help fund the Russian plutonium disposition program. Adequate financing mechanisms have yet to be identified. While the collective efforts of the interested governments will be needed to help solve this problem, the best solution would be to identify some commercially viable and long-lasting activity that would help generate sufficient revenues to enable the Russians to convert safely their excess weapons plutonium stocks to more secure forms. Additional funds also are urgently required for various other efforts to reorient the vast Russian nuclear infrastructure to peaceful purposes and to upgrade significantly the protective measures that apply to vast stocks of sensitive nuclear materials.

Concurrently, there is an urgent need to address the problems some countries are facing in managing their nuclear fuel cycles. Some countries in Asia and in Europe are looking for new options for managing the spent fuel inventories and nuclear wastes being generated in their nuclear power programs. This has been a major preoccupation in South Korea and Taiwan, and Japan is giving increased attention to the interim storage of spent nuclear fuel to supplement its reprocessing programs. In this regard and at the commercial level, proposals are now being put forward to provide some countries with additional constructive options for spent fuel and waste management coupled with revenue generation that would help meet the compelling financial demands associated with reducing the nuclear dangers in Russia. Specifically, several proposals recently have been made that international spent fuel storage facilities or nuclear waste repositories should be established in Russia or in some alternative locations (notably in Australia) with the idea that a substantial fraction of the profits might be devoted to assisting Russia in securing nuclear materials, including plans for converting the tons of excess Russian weapons plutonium to the so-called spent fuel standard. Thus, conceivably, the international community may have a useful opportunity to advance critically important arms control and nonproliferation objectives while it also assists some countries in executing their national nuclear power programs. The identification of improved and more cost-effective opportunities for managing spent fuel and nuclear wastes could be crucial to ensure that existing nuclear power stations remain commercially viable.

Concept

The concept of trying to establish, as a supplement to national programs, some international or regional facilities for the storage or disposal of spent fuel or nuclear wastes has held an attraction for many years. This stems from a recognition that several smaller countries might be unable to establish geologic repositories or long-term storage sites for spent fuel or nuclear wastes on their own, that some countries in Asia and Europe are facing formidable obstacles in arranging politically acceptable solutions for managing their spent fuel or nuclear wastes, and that economies of scale may make it attractive for states to combine their efforts rather than pursue entirely independent national programs. In addition, the establishment of international spent fuel storage or disposal facilities could promote nonproliferation objectives by introducing greater transparency into some nuclear programs and by limiting the buildup of spent fuel inventories at many different locations that will have to be subjected to international safeguards.

Despite these attractions, no credible proposal to establish an international storage site or repository has ever materialized for several reasons. For example, countries are reluctant to become the host for storing or disposing of spent fuel or nuclear wastes received from other countries. Also, it is difficult to get some supplier states, like the United States, to reach agreement with some consuming states on the merits of establishing such facilities because of difference over fuel cycle policies. Nonetheless, pressures are developing in some countries to explore new solutions. It is conceivable that there could now be a greater receptivity to establishing new international facilities, especially if they are perceived as also supporting important arms control objectives.

Proposals

Recently, there have been a number of international proposals calling for international storage of nuclear spent fuel or an international repository of radioactive waste.

Summary of Various Proposals

In November 1998 the IAEA held an international symposium on the storage of spent fuel from power reactors; some of the papers presented were specifically dedicated to the need for international or regional storage or disposal facilities. It was suggested that all types of spent fuel from various types of power and research reactors should be considered for storage in such regional facilities. According to statistics, the annual spent fuel from all types of power reactors amounted to approximately 10,500 tons of heavy metal in 1997. The total cumulative amount at the end of 1997 from all over the world was approximately 200,000 tons. It may surpass 340,000 tons by the year 2010. Approximately 130,000 tons of spent fuel are presently being stored in at-reactor or away-from-reactor storage facilities, awaiting either reprocessing or final disposal; and 70,000 tons have already been reprocessed. Right now the quantity of accumulated spent fuel is 20 times the present total annual reprocessing capacity. Considering that the reprocessing capacity will not significantly increase in the near future, the necessity of interim storage is indisputable, and the need to provide such storage facilities may lead to an attractive commercial market.

Wolf Häfele and Chauncey Starr have advocated the concept of an Internationally Monitored Retrievable Storage System (IMRSS). The objective of the IMRSS is to establish an international framework for a monitored retrievable storage system for spent fuel generated from power stations and also for surplus plutonium. The concept is designed to place the spent fuel and surplus plutonium of each country under international control through the IMRSS, which is to be operated on a commercial basis. It purports to place spent fuel and plutonium into storage within a highly transparent framework until such time as the commercial use of plutonium becomes more attractive or until further decisions are made.

Boris Nikipelov and others from MINATOM have disclosed the idea that Russia is considering the possibility of storing spent fuel shipped from other countries. This proposal was designed to be a unique form of international cooperation. In compliance with the current Russian federal law that prohibits the receipt of spent fuel as a waste, the MINATOM proposal is designed eventually to provide reprocessing services. What is crucial in the Russian proposal is whether Russia is able to provide a spent fuel storage service without any commitment to reprocessing or preservation of the option to reprocess. Actually, both MINATOM and the Russian Parliament are now working together to change the law to allow the receipt of spent fuel for commercial purposes.

Still another recent proposal by Matthew Bunn, Neil Numark, and Tatujiro Suzuki presented some very specific ideas to help resolve both the financial difficulty associated with disposition of surplus nuclear materials in Russia and the shortage of domestic storage capacity for Japan's and possibly other Asian countries' spent fuel. The concept, in essence, would call for a Japanese-Russian government-to-government agreement (with support from additional partners as desired by the principal two) on siting cooperative facilities in the Russian Far East. The facilities would address the global-security goal of fostering the disposition of weapons plutonium while in parallel providing much-needed storage capacity for spent nuclear fuel from commercial Japanese power reactors as well as retired Russian naval propulsion reactors.

Under the proposal, if acceptable terms could be agreed between the Japanese government and Japanese private electric utilities, such utilities would be the principal beneficiaries of new facilities to be built in Russia because they would secure (a) low-cost contracts to store their spent fuel at the facility for a specified period, perhaps as long as 50 years; and (b) low-cost contracts for supply of MOX (mixed oxide) fuel using Russian excess weapons plutonium. The revenues generated from Russia's provision of these commercial services to Japan's and possibly other Asian utilities would contribute substantially to financing the capital costs of both the MOX plant and the storage facility to be built in Russia and could also provide additional funds for Russia and the population of the affected region.

Thomas B. Cochran and Christopher E. Paine of the Natural Resources Defense Council in the United States have also made a proposal about storing spent fuel from Asia and Europe in Russia. The idea is to establish an international storage facility for spent fuel in Russia in a collaborative arrangement between a private U.S.-based "nonproliferation trust" and MINATOM. Under this proposal, the trust would take title to foreign commercial spent fuel that would be moved to Russia for storage and that would contain up to 50 tons of fissile plutonium in return for a fee approximately equal to the avoided cost of spent fuel storage, reprocessing, and high-level waste disposal.

Under the proposal, some of the proceeds and profits would be employed to foster the improvement of the security of sensitive nuclear materials and facilities in Russia and to address environmental issues. In addition, the trust would lease and presumably would have under its control for a specified period (20 to 30 years) up to 50 tons of excess weapons plutonium that the trust would obtain from MINATOM. The fate of this plutonium and the arrangement for its disposition would have to be agreed to between the trust and MINATOM.

At the CSIS meeting held on December 4, 1998, Atsuyuki Suzuki proposed another scheme designed to provide potentially long-lasting funds for Russia and to create benefits for both reduction of nuclear arms and future civil use of nuclear energy. Specifically, Russia would provide spent fuel storage services to other countries, and Japan and other countries would ship their own spent fuels to Russia to store for a definite period of time—50 to 100 years, for example. A rough estimate shown in the proposal is that a potential market of such spent fuel storage would not be less than 500 tons per year, even if it is limited to the East Asian market only (i.e., Japan, South Korea, and Taiwan). Provided that the 20-year cumulative amounts of spent fuel of those countries can be stored in the host country, Russia would receive at least $1 billion in 20 years. This assumes that the storage costs approximately $100,000 per ton, which is a very rough estimate based on the current world average cost of spent fuel storage.

Under this proposal, the location of such a spent fuel storage facility would be at a MINATOM complex, where the required infrastructure is available. The construction and operation of the facility would provide the site with a significant amount of revenue, which could substantially help economic development for transparent denuclearization, transparent civilian use, transparent consolidation and storage of nuclear materials, and transparent safety and security.

Finally, David Pentz has proposed another new regime to establish an international repository for nuclear wastes and spent fuel. The idea, which is named the Pangea concept, calls for establishing a permanent geological disposal facility in Australia, where the geology and biosphere conditions show such simplicity coupled with robustness that they can demonstrate that the facility will fully meet the highest international safety standards and safeguard requirements. One option under this proposal is that the geological disposal facility would be devoted to holding not only spent fuel or highly radioactive wastes arising from civil use of nuclear power, but also surplus weapons material from the United States and Russia that has either been irradiated as MOX fuel for light water reactors or immobilized within technically safe media.

The Pangea concept is particularly noteworthy in that it provides a state, like Australia, with the opportunity to play an unprecedented role in hosting an international repository that also would serve to enhance nonproliferation and arms control objectives. The principle of trust and symmetry between nuclear weapon states coupled with the presence of an international repository in a nonweapon state would open up new considerations to strengthen nuclear security and nonproliferation.

Conditions for Any Successful Proposal

Of course, it is assumed that no proposal to establish a regional or international facility for the storage of spent fuel or disposal of radioactive waste will be seriously pursued unless it is commercially viable for the host country and some consumer nations. (The CSIS group believes that the proponents of some of the new concepts being discussed should be encouraged to explore creative new funding mechanisms, including access to resources of the World Bank.) Beyond this, the establishment of any such facilities will have to meet some very basic tests—and overcome some significant hurdles—if they are to be truly viable.

- Clearly, an international spent fuel storage or waste repository will need to be acceptable to the government and population of the host country and, most likely, some of its neighbors. There will have to be high local confidence that the benefits will offset any costs and that the proposed facility will be constructed and operated under conditions that will offer no significant risks to the public health and safety and to the environment.
- The challenges associated with establishing an international permanent repository for spent fuel and nuclear wastes as opposed to an interim spent fuel storage facility will be somewhat different. To be viable, a repository will have to be in a suitable, technically acceptable geologic location, and the host will have to be prepared to accept wastes and spent fuel from other countries on a permanent basis. A proposal by a state to receive and store spent fuel from other states on an interim basis entails less of a permanent obligation but leaves for later determination and negotiation what happens to the spent fuel at the end of the agreed and defined storage period.
- There also will have to be high confidence by respected scientific groups that the project is well conceived and designed from a technical, scientific, and engineering perspective. The project also will have to be well defined to ensure the timely implementation of the venture, ranging from the presence of the necessary roads, port facilities, and rail lines to the presence of adequate technical and engineering personnel.
- The host state will have to have the requisite regulatory infrastructure in place to review and oversee the venture, and there will have to be high confidence that the project will be able to meet the applicable laws, rules, and regulations that are in effect. As an example, sending spent fuel to Russia for storage or permanent disposal will require the processing of a new law through the Duma.
- In addition, because spent nuclear fuel is sensitive from a nonproliferation perspective, there will have to be high confidence in the political and institutional stability of the host country on the part of all other states whose concurrence

will be required to enable the venture to move forward. This, obviously, will include the prospective customers as well as any supplier states whose approvals might be required to permit transfers of material. Clearly, the host state will have to be politically acceptable to all interested parties.

- There will have to be a basic compatibility of views on the nonproliferation policies and conditions that should apply to the materials involved among the host country, the transferring or customer states, and the supplier states that may have prior consent rights over the nuclear materials involved. For example, under some proposals, it is contemplated that some special nuclear materials of U.S. origin that are now subject to U.S. consent rights will be proposed for transfer to possible storage or disposal facilities in Russia or other countries. Under U.S. law and the terms of all applicable agreements, the U.S. government would have to give its prior approval to any such transfer, and it could not be expected to approve any such transfers unless it was fully satisfied with the nonproliferation conditions, including the safeguards and physical arrangements that would apply to the project.
- In addition, under U.S. law, it would not be possible for the United States to approve any transfers of U.S.-obligated spent fuel from a client state to an international storage site in another country unless the United States has in force a suitable agreement for cooperation with the host government. The United States has no such agreement for cooperation in force with Russia although it does have a suitable agreement in force with Australia. More broadly, if any international facilities receiving U.S.-origin spent fuel materialize, it can be expected that the United States would have a strong interest in the nonproliferation arrangements that would apply to such facilities and might wish to constructively participate in their development.

Benefits

Some of the international nuclear security, safety, and environment issues conceivably could be ameliorated through the prudent establishment of an international storage facility for spent fuel or an international repository for nuclear wastes. First, significant challenges need to be overcome to help ensure that sensitive nuclear materials and facilities in Russia are placed under far better control. Second, within the civil nuclear sector, there is a need to identify and develop additional options that will enable some states with rapidly accumulating inventories of spent fuel to better manage these inventories under conditions that will be fully compatible with nonproliferation and environmental values. Some proposals for establishing international spent fuel and/or waste facilities hold out hope of constructively contributing to both of these objectives. The facilities would provide funds for improving the security of nuclear materials (including excess weapons plutonium in Russia) on one hand while they would be able to meet international high standards for public health protection and accountable management of materials and add a barrier to nonpeaceful diversion of spent fuel on the other.

In particular, a potentially significant benefit is associated with transparency; this could be achieved in two different ways. First, the nuclear problem in North Korea, directly related to the question of plutonium extracted from nonpeaceful reprocessing of spent fuel, was compounded by the lack of transparency and accountability. This suggests the potential advantage of an international spent fuel storage facility that is multilaterally managed and subjected to international safeguards. An international storage facility or regime could also significantly contribute to mutual confidence building. Second, if the international storage facility is built in Russia, it could probably provide the opportunity to improve transparency, which is particularly necessary in that country.

A new international storage facility would help promote constructive collaboration and partnership among the participants in various other ways. One postulated goal for the various international storage or repository schemes that have been proposed is to generate revenues to assist the transition of the Russian complex. If the facility was located in Russia itself, it would provide additional transparency about the Russian situation and would provide useful jobs for Russian nationals who would otherwise find it difficult to be engaged in constructive peaceful activities.

One of the most crucial issues facing the civil nuclear industries in many countries is the problem of managing spent fuel, including interim storage and geological disposal of highly radioactive waste. A number of small countries with small nuclear power programs face serious problems of how to manage the extended interim storage and disposal of their spent nuclear fuel. For these countries it is too expensive to construct their own away-from-reactor storage facilities and/or geological repositories because of the limited amount of spent fuel and, in some cases, geographic constraints. Proposals to establish regional or international interim storage facilities and/or repositories would possibly resolve the problem.

An international facility would also provide more flexible choices for participating partners in implementing their programs of spent fuel management. If a consuming country has acceptable arrangements with an international storage facility, the country would have the opportunity to ease the shortage it might face in domestic storage capacity. If a client country considered that its future decisions on reprocessing or direct disposal should be flexible to follow the future global energy situation, it could buy time until the future use of plutonium becomes better clarified by placing spent fuel into storage within an international framework. It might also provide for sound and healthy competition if the regime has more than one international storage facility. There are potentially enormous demands for international storage of spent fuel; hence, it would make economic sense to have more than one service-supplying facility.

According to IAEA estimates, the total world cumulative amount of spent fuel might surpass 340,000 tons by the year 2010. If 100,000 tons of spent fuel, which is less than one-third of the total amount by the year 2010, was stored internationally by a host state, the host country could receive approximately $10 billion. This assumes storage costs of approximately $100,000 per ton, which is a very rough estimate of the current world average cost of spent fuel storage applied for a much smaller facility such as at-reactor storage. The storage cost decreases significantly with the increase of scale, and a large portion of the $10 billion revenue would be profit available for funding the Russian denuclearization program. If the international facility is a final repository, the revenue might be even larger. Although the cost for final disposal strongly depends on various factors including the geological environment and the capacity of the repository, a rough estimate can be provided based on costs in the United States, where the total cost required for the Yucca Mountain repository for disposing of approximately 70,000 tons of spent fuel is estimated to be more than $36.6 billion. This suggests that a revenue of tens of billions of dollars could be expected from an international repository.

On the whole, it would appear that international or regional facilities for storing or otherwise disposing of spent nuclear fuel and/or nuclear waste could offer opportunities for producing significant revenues. Such revenues could help finance the transition of the Russian nuclear establishment if the profit raised by the economics of scale inherent in cost sharing is properly managed so that it may be specifically dedicated to such assistance.

Obstacles

A number of significant obstacles and hurdles in establishing an international program of spent fuel storage or radioactive waste disposal facility would have to be overcome. There would be at least two types of concerns associated with the proposed establishment of an international facility.

First, if the facility was to be located in Russia, many people might be concerned about the uncertainties associated with the future of the Russian political and social systems—whether Russia has a legal and administrative structure that is robust and attractive enough to outside investors to accommodate an international facility. This is particularly questionable with respect to tax and liability laws. Given the considerable political instability in Russia, there might be substantial doubts whether the prospective partners would be prepared to transfer in the near future their spent fuel to Russia for storage in the facility, even if there were strong economic motivations to pursue such an option. It seems likely that several participating partners may be prepared to pursue such an international storage venture only if the host country is perceived to be politically stable and they have confidence that the U.S. government will be truly supportive of the proposed venture.

A second type of obstacle could be connected with the movement of spent fuel that contains a significant amount of plutonium. One of the prerequisites for an international scheme to store spent fuel will be a consensus among participating countries that the spent fuel will not be reprocessed but only be stored on an interim basis or disposed of as an industrial waste. It is not clear whether such conditions would be acceptable to Russia, given that MINATOM has an avowed interest in providing reprocessing services. Current Russian law has to be changed to allow non-Russian spent fuel to be received in that country for a protracted period. Efforts are being launched to amend this law, and there is a good chance that the legislation may be changed to permit the import of foreign spent fuel for storage.

Various environmental groups will oppose any change that would permit the import of foreign spent fuel into Russia, however.

The movement of spent fuel will pose some important issues for United States. As is well known, much of the spent fuel in Japan, Taiwan, and South Korea, for instance, is of U.S. origin and cannot be transferred to a third-party state in the absence of a bilateral nuclear cooperation agreement between the host country and the United States. This is the case if a facility is built in Russia. Therefore, it goes without saying that the establishment of a stable nuclear relationship between the United States and Russia, including the conclusion of a U.S.-Russian agreement for cooperation, will be a prerequisite for implementing the international storage facility in Russia.

Some may question the merits or wisdom of mixing up or combining the challenge of finding a solution to the spent fuel handling problem with the issue of deciding how best to finance the Russian program aimed at the transition of the nuclear complex, including the disposition of excess weapons plutonium. Some may simply wonder why the United States should take a supportive role in helping to build an international storage facility in a certain country so that it may receive spent fuel being generated in other countries. This argument is quite understandable, and it will be necessary for the participating countries or industries to demonstrate that this combination of the two objectives will foster better progress than a separation of the two objectives.

Recommendations

These various considerations strongly suggest to our CSIS working group that several of the concepts now under study for establishing international spent nuclear fuel storage or radioactive waste repositories merit further consideration. The U.S. executive branch and the Congress should be prepared to give encouragement to the further development and elaboration of these ideas—recognizing that to be credible they will have to meet some crucial tests and preconditions. In the first instance, it will be up to the prospective investors to determine whether they appear to have technically, politically, and financially viable proposals in hand that also promise to meet the applicable public policy, public acceptance, regulatory, and nonproliferation conditions.

The U.S. executive branch should be prepared, if so requested, to outline the kinds of criteria a project would have to meet to enable special nuclear materials that are subject to U.S. consent rights to be transferred from various countries to host sites.

The U.S. government also should be prepared to engage in constructive exploratory conversations as to how prospective projects might be shaped to enhance the prospects of U.S. support. In this regard, several members of the CSIS group are of the strong view that within the current international climate, a proposed international storage or disposal facility will stand the greatest chance of success in garnering international and U.S. support if it demonstrably makes a contribution to the plutonium disposition program in Russia and also helps nations practically deal with their growing inventories of spent fuel or nuclear wastes. Still, while it is desirable to explore the feasibility of regional storage or repository proposals aimed, in part, at producing revenues for transitioning the Russian complex, this should not be at the expense of vigorous efforts to develop other funding and assistance mechanisms that will be directed at this very same objective.

It will be important to conclude a new U.S.-Russian agreement for cooperation to permit any U.S.-origin spent fuel to be transferred to Russia for long-term interim storage or final disposal. It is recommended that early discussions be initiated at the governmental level between U.S. and Russian representatives on Russia's interests in this area, on the political conditions that would have to be met to enable the two sides to negotiate such an agreement, and on the likely nonproliferation conditions that could have to be included in any such agreement for cooperation.

The U.S. government, and notably DOE, should also underscore its willingness to share with all of the interested parties the full benefits of the technologies that have been developed by the U.S. government on the management and disposition of spent fuel and nuclear wastes.

In addition, the members of our working group are of the strong view that the ability of the United States to remain a credible collaborator with other countries in searches for new institutional solutions for dealing with nuclear wastes or spent fuel will greatly depend on the U.S. ability to now bring its own nuclear waste program to some form of closure—under conditions that are broadly acceptable within the United States.

Finally, the concepts being discussed largely focus on the merits of offering new options for spent fuel management to countries like South Korea, Taiwan, and Japan, as well as some nations in Western Europe. Clearly, the attitudes these nations have toward any storage or repository proposals as supplements to their national programs will be crucial to their success.

It was pointed out to our group that Japan is in a special position to help advance some of the regional storage concepts that have been proposed because Japan is running out of domestic capacity for spent fuel storage at reactors and urgently needs to assess and develop alternatives. More important, as one of the most advanced nuclear nonweapons states, Japan has a strong historical interest in nuclear disarmament. In addition, Japan will be hosting the G-8 Summit in the year 2000. Thus, the Japanese government should be encouraged to take a leadership role at the summit in discussing the ways arms control and nonproliferation goals can be advanced, and it should be encouraged to initiate an early dialogue with the United States and the other G-8 countries on that subject.

CHAPTER 3

Commercializing the Excess Nuclear Defense Infrastructure

Task Force III

WHAT ARE THE POSSIBLE PATHS TO COMMERCIALIZING EXCESS NUCLEAR DEFENSE INFRASTRUCTURE?

This task force report addresses the issues arising from the Cold War nuclear legacy of the excess nuclear defense infrastructure. These issues include:

- Conversion of excess defense nuclear facilities and other assets to transparent, nondefense use,
- Treatment of weapons-grade nuclear material,
- Decommissioning of nuclear submarines,
- Treatment of spent nuclear fuel,
- Environmental cleanup of contaminated military nuclear sites, and
- Deployment of staff and skills in defense and research establishments such as the Russian closed cities.

The focus of the task force is not on the commercial handling of the Western excess defense infrastructure that is paid for by Western governments, where the commercial paths are reasonably clear-cut. In a Western market economy, with adequate employment and mobility of labor, the conversion of a defense to civilian infrastructure—although not necessarily easy—is relatively well charted.

This report focuses on the vast problems exemplified by the Russian nuclear defense infrastructure, where economic difficulties and politicoeconomic ramifications complicate the defense conversion and commercial paths. Russia does not yet have a developed market economy or mobility of labor, but neither does it have a command economy. In an old-style Soviet economy the defense infrastructure could have been ordered to make civil products. The market might not have wanted these, but the conversion process could have been carried out with an artificially maintained level of employment. The problem with defense conversion in Russia today is linked to the much wider problem of commercializing a developing market economy in Russia and cannot be viewed in isolation; the development of a commercial environment that is conducive and safe for foreign investment is a critical national issue.

In the context of this report, it is important to differentiate what is meant by the terms "defense conversion" and "commercialization." Defense conversion is that activity that must be led by governments in order to make the defense activity fit for civilian purposes to meet current and anticipated needs. It is implicit that this will require significant financial subsidies to achieve its goals. Commercialization refers to activities that may result from defense conversion where new products or services are able to compete successfully in a free market and their success, or otherwise, is determined by their sustained profitability. This will also probably require government subsidies, at least initially, but the eventual success of any commercialization efforts will ultimately depend on their profitability. "Excess" materials are defined as everything not needed to support Russia's nuclear weapon stewardship and naval propulsion infrastructures.

Although commercial companies can assist in the defense conversion process, it is essential that governments first establish clear, targeted, and achievable strategic objectives within which industry can operate. In the case of the Russian nuclear defense conversion, the priorities for Western governments are to ensure that:

- All nuclear usable materials are kept under safe and secure conditions,
- Nuclear weapons scientists and technologies are not redeployed to internationally undesirable activities, and
- Nuclear environmental problems of actual or potential international significance are adequately addressed.

It is true that Western resources have focused on these three key areas, but it is also the case that there has been inadequate coordination of international effort and a lack of political leadership in some areas to achieve the necessary and timely progress that the situation requires. Similarly, governments have not yet made the necessary changes to national and international laws and regulations that are prerequisites for successful commercialization, as this report highlights. Task Force III hopes that the recommendations of the report will go some way to helping redefine the priorities and actions necessary to begin to attract additional foreign investment into the Russian defense conversion process. This is key to success although it is acknowledged that the path will be difficult and take many years to accomplish.

The health of the Western nuclear industry is a relevant factor in this whole debate. A reinvigoration of the nuclear power market (particularly if concerns about climatic changes persuade governments to look favorably on nuclear energy sources) and the international consolidation of the Western nuclear industry will be important for enhancing the nuclear industry's capability to help promote the priorities noted above. Conversely, a declining Western industry based on traditional, national models faced with unnecessarily restrictive trade barriers will have less freedom of maneuver to assist in these important tasks.

The task force addressed several issues:

- Can excess nuclear defense materials be treated on a straight commercial basis?
- Is the condition of the Russian economy likely to make commercial paths feasible in the foreseeable future?
- Does Western environmental grant/loan aid make nuclear environmental cleanup a viable commercial proposition?
- Is Western direct aid for military decommissioning programs commercially viable?
- What are the wider politicoeconomic benefits of providing employment for skilled staff in Russian defense establishments, and what are the commercial options for helping these establishments convert to civil tasking?
- What are the wider and longer-term commercial benefits of nuclear environmental cleanup, such as protecting the Norwegian fishing industry or stimulating inward investment to Northwest Russia, thereby helping to develop a regional market economy?
- Given these considerations, should Western governments and companies take a longer-term politicoeconomic view of these issues instead of a short-term commercial view?

U.S. Perspectives

As stated in the introduction, this report focuses not on U.S. perspectives but on the complex Russian perspectives although there are important lessons to be learned from the U.S. experience. For the same reasons this report does not address West European or Chinese perspectives, which would be similar to U.S. or Russian experiences on a smaller scale.

Before we address the Russian perspectives, however, it is useful to give examples of the U.S. experience in commercializing the excess nuclear defense industry. A key aspect of this conversion was an essentially flat budget for nuclear defense activities after the Cold War (in the period 1992–1998); while there was a reduction in nuclear-related production activities, there was an increase in spending for cleanup and conversion that offset this decrease in production spending.

We summarize the major initiatives below.

Consolidation

DOE is downsizing its weapons complex, which will involve moving functions from many sites to a few sites. Thus, for instance, Rocky Flats plutonium pit recycling is being relocated to Savannah River. Rocky Flats is being decommissioned and will be shut down after decommissioning and decontamination (D&D) are completed. All plutonium materials of good quality (including pits) will be shipped to Savannah River. All plutonium-contaminated waste will be shipped to the Waste Isolation Pilot Plant (WIPP), a repository for transuranic wastes that just recently opened in

Carlsbad, New Mexico. Some assets are being converted or added to support the following mission areas—stockpile stewardship, navy propulsion fuel cycle, arms control and reduction, proliferation prevention, material treatment consolidation and disposition, energy research and dismantlement, and environmental cleanup.

Disposition of Excess Weapons Materials

Plutonium. This program is dual track. It seeks to (1) manufacture MOX from plutonium in excess weapons and burn in commercial nuclear power plants, and (2) immobilize plutonium in excess weapons in ceramic matrices for direct disposal. The program will progress at a pace equal to the Russian plutonium disposition program Thus, U.S. progress will be tied to comparable Russian progress.

HEU. This material will be blended down for manufacture of commercial LEU fuel.

Worker Transition

DOE has a program in place that provides worker retraining and preferential hiring of this "old" work force for any new DOE projects on the site.

Environmental Cleanup

Cleanup is the main activity at most of the former weapons production sites. DOE funding in environmental cleanup runs at approximately $6 billion per annum and is a major source of jobs at these various sites.

Reindustrialization

Contractors at the DOE sites are given incentives to assist the communities in developing new business to replace the weapons missions. One example of this is the East Tennessee Technology Park, which involves the demolition of the Oak Ridge gaseous diffusion plants; recycling of the metals taken from within the plants for beneficial reuse; and use of the emptied, decontaminated structures or sites for new industrial purposes.

Defense Conversion and Commercialization

DOE has provided enrichment services for both U.S. government and commercial nuclear markets. Early components of this infrastructure are being decommissioned and dismantled. The remaining facilities and supporting infrastructure are being commercialized by transitioning the assets to the U.S. Enrichment Corporation, but this will take several years to accomplish. The complex is also supporting the proliferation prevention and arms control missions of the U.S. government.

Summary

Consolidation, disposition, commercial use, retraining, reindustrialization, reuse, and downsizing of the DOE weapons complex are all parts of the U.S. DOE program to address the redundancies in existence throughout the DOE complex. All these programs are extremely costly—in the tens or even hundreds of billions of dollars over the lifetime of the projects. It is important also to recognize the difficulties experienced in achieving the successful commercialization of the U.S. program in a thriving economy and free market with very few institutional barriers. The survival rate of new businesses formed as a result of defense conversion activities is approximately 20 percent after five years and this may be due, in part, to a lack of commercial awareness and marketing skills among the nuclear defense communities. It is probably also due to a significant difference in perception between the apparent value of nuclear-related assets and their true commercial worth.

One example of this is a Florida-based facility (land, property, and infrastructure) that cost over $300 million but that was valued by the commercial market at just $2.8 million. This significant reduction in value occurred because all that was left was real estate—that is, there was no way to transition the site activities to a service needed in the civilian market. This analysis may well indicate that the goal of commercialization in Russia will require significant and fundamental changes if it is to have any chance of success in the future. It also draws attention to the need for objective and dispassionate analysis of commercial proposals and market opportunities before significant investment occurs.

Russian Perspectives

We can assume that the dimensions of handling the Russian nuclear defense excess are of the same magnitude as in the United States—tens or hundreds of billions of dollars. The condition of the Russian economy for at least the next 10 years, however, means that direct Russian state funding for tackling the problems of the nuclear defense excess will be limited. For example, the Russian Federation program of October 23, 1995, for the "management, reclamation and interment of radioactive waste and spent nuclear materials for the period 1996–2005" estimated the cost of radioactive waste management (arising from mining, processing, power stations, reprocessing, weapons, ships, scientific research work, and underground explosions) at 8,700 billion rubles. Of this, 4,700 billion rubles were to come from the Russian federal budget and 4,000 billion from Russian commercial enterprise (1995 prices). These figure are debatable, but give an idea of local estimates—although the real cost is likely to be considerably higher.

The Russian federal budget and commercial sector are not capable of financing programs at that level. The Russian gross national product (GNP) fell by 5 percent in 1998 and a further 4 percent in 1999. Investments in Russia fell 40 percent in the first quarter of 1999. The total indebtedness of the Russian economy is 146 billion rubles against a federal budget of 20 billion rubles in 1999. It is therefore clear that wider Russian nuclear defense conversion programs cannot be financed without very substantial international support and the generation, by MINATOM, of significant export markets worth billions of dollars.

Nevertheless, the condition of the Russian excess nuclear defense material poses environmental, security, and terrorism threats that the international community cannot afford to ignore while waiting for the developing Russian market economy to improve. This is the crux of the whole problem. Unless fundamental changes in the Russian financial, legal, and regulatory systems are made, Western governments and commercial organizations will not have the confidence to make the scale of financial investment that is necessary to provide the West with the political assurances that it needs.

Several barriers are unique to or much more pronounced in the Russian defense infrastructure:

- There has been no requirement in the past to segregate military from civil (peaceful) nuclear activities in either storage or processing of nuclear materials. If Russia has ambitions to take maximum advantage of international nuclear fuel service opportunities, it will be necessary to segregate these activities in order to provide the necessary assurances to foreign governments that peaceful use materials and technologies are not being diverted from their intended, peaceful use. In the short term, and in common with nuclear facilities in other nuclear weapon states, this could be achieved through the temporal separation of these activities in common facilities (so-called mixed plants) so long as civil activities are fully transparent and subject to international verification. In the longer term, it will be necessary to build or restructure such plants so that spatial separation is also achieved, perhaps by bringing all nuclear activities and materials on the site under international safeguards. This should be achievable, given that the Russian defense conversion process should consolidate defense activities onto a smaller number of sites, including perhaps Arzamas-16 and Chelyabinsk-70, thus freeing up sites such as Chelyabinsk-65, Tomsk-7, and Krasnoyarsk for civil nuclear services. This will require an integrated defense conversion plan and an effective nuclear material stewardship program to be established in Russia that also address the very important cultural changes (i.e., attitudes to international verification) that will be entailed.
- It is inevitable, in our view, that a significant number of sites and facilities that currently exist in Russia will have to close because there is no realistic prospect of commercializing their activities. For that reason, MINATOM should focus its efforts and resources on those sites and institutes with future potential, however unpalatable this may be.
- Currently there are enormous barriers to investment caused by the institutional attitude toward the closed nuclear cities. These are manifest in the difficulties experienced in accessing and working on the sites. Until the Russian government recognizes that steps must be taken to make the sites more accessible, consistent with Russian national security, it is difficult to see how foreign commercial investment can proceed.
- Russian tax laws, customs procedures, liability issues, and assurances over repatriation of foreign investment all need radical reform.
- The Russian nuclear industry has only limited access to the international market. If Russia wishes to exploit global nuclear markets in full, it is essential that bilateral trade agreements be negotiated with other countries, including the United States. For example, if Russia decides to offer spent fuel storage or waste

disposal services to foreign utilities, it will probably have to conclude peaceful nuclear cooperation agreements that will require, among other things, the acceptance of international safeguards. An agreement with the United States is also important because the United States has a significant although decreasing global influence on nuclear trade because of its bilateral trade agreements, which include nonproliferation controls; and U.S. consent would be required for the transfer of U.S. controlled nuclear material into Russia.

Against this background, we summarize major Russian initiatives where there are defense conversion activity and possible commercial opportunities.

Treatment of Russian Weapons-Grade Nuclear Material and International Spent Fuel Management Services

HEU DEAL. On March 24, 1999, the Western companies Cameco, Cogema, and Nukem signed a contract with Russia's Techsnabexport (Tenex) giving them the option to purchase up to 100,000 metric tons of uranium from down-blended Russian warhead HEU. This contract, which has the blessing of the U.S. and Russian governments, is for 15 years, the expected remaining life of the U.S.-Russian HEU agreement. The feed component from the down-blending of 30 metric tons of HEU each year produces LEU, with a feed component of 9,000 metric tons of uranium as UF6 or 24,000 pounds as U_3O_8 equivalent.

A key element of the government-to-government (DOE-MINATOM) agreement is the creation of uranium stockpiles in the United States and Russia. DOE paid Russia $325 million to purchase the 1997 and 1998 uranium deliveries associated with the HEU deal and will hold these and a similar amount of DOE's own uranium—a total stock of 22,000 metric tons—off the market for 10 years to reduce oversupply in the market. Uranium, which Western companies do not buy, will be shipped back to Russia, where a comparable stockpile will be created. The agreement permits the U.S. stockpile to be reduced to ensure the reliability of deliveries under the commercial agreement. The deal was designed to help stabilize the uranium market and to support higher prices. At current conversion rates the time for total conversion of the Russian and U.S. HEU surpluses extends to 2051. Proposals have been made by Pangea in Australia for speeding up the conversion rates by a G-8 leasing concept with access to an international repository. Task Force II has reviewed this and other schemes.

PLUTONIUM. The major obstacle to the disposition of excess Russian weapons plutonium is MINATOM's inability to finance such a program. (The costs of converting the plutonium, fabricating MOX, and completing the modification of Russian reactors could cost billions of dollars.) No definitive scheme for financing this program has emerged, and this could impair the credibility of this important enterprise. Proposals to use the revenues from additional sales of LEU down-blended from Russian HEU (beyond the 500 tons in the U.S.-Russian HEU deal) have been made to help finance a MOX plant in Russia for the disposition of excess Russian weapons plutonium in reactors. Proposals have also been made for the establishment of an international spent fuel storage facility in Russia and for the profits from such an enterprise to be used to finance a MOX plant in the Russian

Far East from where MOX fuel would be shipped to Japan for irradiation in Japanese reactors. Such proposals face formidable obstacles and would require the support of the U.S. government because prior consent rights are attached to most of the fuel in question. At a minimum, this would require the negotiation of a U.S.-Russian peaceful nuclear cooperation agreement and would have to conform to international verification and security standards.

To date, the United States has declined to negotiate such an agreement because of its objections to Russian assistance to the Iranian nuclear program. There has also been strong opposition from international environmental organizations, which argue that if Russia cannot cope with its own nuclear materials, how can it safely store additional foreign spent fuel? These issues and possibilities are being considered by Task Force II but may prove to be the key for attracting the necessary foreign investment to help the Russian nuclear industry restructure.

It has also been proposed that Russia could extend its international reprocessing services from either Mayak or Krasnoyarsk (RT2). Although this could lead to the development of significant numbers of new jobs at Russian facilities and the construction of modern, high-technology facilities to handle existing and future radioactive waste streams and residues (akin to the British experience at British Nuclear Fuels Limited [BNFL], Sellafield), there is likely to be significant political objection from the United States. The current U.S. policy is not to encourage commercial reprocessing and the use of plutonium on nonproliferation grounds.

Russian Submarine Decommissioning, Reprocessing of Spent Fuel, and Waste Management and Cleanup of Contaminated Defense Sites

Following the Chernobyl disaster the international community focused initially on the environmental consequences of civil nuclear reactor safety. G-7 and European Union (EU) funds of the order of $500 million plus are being channeled through the European Bank for Reconstruction and Development's Nuclear Safety Account (EBRD/NSA) and the Tacis program (an EU program to aid newly independent states in the transition to market economies and the reinforcement of democracy) for tackling the environmental problems posed by Chernobyl and reactor safety throughout the former Soviet Union. These programs do not address the environmental problems in Northwest Russia and the Nordic region that are posed by spent fuel from the more than 100 decommissioned submarines from the Russian Northern Fleet.

In recent years there has been increased international attention on this area, focused initially through a joint Russian-Norwegian program. Norway, being directly threatened by the environmental consequences, for example, to its fishing industry, has contributed more than $50 million to remedial activity. Norway's Nordic neighbors, Sweden and Finland, and the United States (because of the Arctic connection) have also contributed funds—less than Norway, but still significant. Most recently, the United Kingdom committed more than £3 million. The Directorate General XI (DGXI) for the Environment, Nuclear Safety, and Civil Protection of the European Commission has contributed from a limited funding base (approximately 1 million euros per annum), and there are indications that the Tacis program may now contribute substantially more (approximately 20 million euros). There are also indications that some of the multilateral loan institutions, which have hitherto not lent on nuclear projects, may be considering support on an environmental basis. The IAEA-based Contact Expert Group plays a role in coordinating priorities for this activity and its external funding.

There is parallel development in the Russian Far East, where the Russian Pacific Fleet has generated similar, if smaller, environmental problems—with remedial measures to be funded externally by Pacific neighbors. The Mayak Production Association in the Urals also plays an important role in these environmental programs although it does not have direct access to the Nordic funding base. It is the central Russian reprocessing facility for spent fuel from the Northwest and Far East. In addition, major nuclear accidents there in the 1950s and 1960s have left large areas of contamination, with river chains leading to the Arctic Ocean.

As noted earlier, progress is held back by negotiations with the Russian authorities over access to the (still closed) sites, and over liability, customs, and taxation. The responsibilities for, and ownership of, the nuclear defense conversion programs are the subject of an evolving process among the Russian regional and central authorities and civil and defense authorities. There are Western sensitivities about assisting Russian defense activity (eased by the transfer of the spent fuel responsibility from the navy to MINATOM) and about treating spent fuel in the reprocessing cycle. These issues are gradually being resolved, however.

Thus there is now a significant external grant-loan funding base for nuclear environmental work in the Nordic area (and to a lesser extent in Mayak and the Far East), where Western commercial companies can play an active role in partnership with Russian enterprises. An inevitable consequence of the Western–Russian bureaucratic process is that initial work has tended to consist of a plethora of small-scale studies and consultancies, not substantial projects. However, there is Western and Russian pressure for this to change. Clearly, commercial margins are going to be tight because the Russian enterprises will try to obtain a major share of the funding. This is a matter for hard negotiation and for local business experience. Western companies bring complementary technology and access to external funding. Such joint activity has the strong support of the Russian regional authorities who recognize not only the local environmental advantages but also the commercial advantages for Russian participation in joint projects—and also the longer-term commercial advantages of creating a cleaner business environment that will eventually attract mutually beneficial foreign investment. Such commercial opportunities will not necessarily be in the nuclear sector, which raises the questions of barter or countertrade (e.g., defense shipyards servicing civil ships).

An interesting example of how this regional approach is developing toward Northwest Russia is demonstrated by the establishment in 1998 of the Euronord-Eco Foundation by 11 regional governors from the Nordic region (including Russia, Norway, Finland, and Sweden). Their aim is to develop a regional solution to the environmental and linked economic problems of the Nordic region, thereby increasing both Russian and Western commercial opportunities in an environmentally restored region. A parallel might be drawn with the cooperation between the mayors of Eisenstadt and Sopron on either side of the Austro-Hungarian border in the late 1980s, before the iron curtain came down. The two administrations redeveloped historic economic, commercial, and social ties between the regions, and the central governments on both sides of the iron curtain let them get on with it.

Conversion of Russian Defense Establishments to Commercial Civil Tasking

Another legacy of the end of the Cold War is excess shipyard capacity in the Russian defense sector. Although there are various examples of this, we have chosen as an illustration the Nerpa shipyard in the Kola Peninsula. The shipyard, with a support population of approximately 15,000, has been unable to remain financially viable since its defense work was reduced after the end of the Cold War. The U.S. Department of Defense Cooperative Threat Reduction (CTR) program, which pays for the decommissioning of two and one-half nuclear submarines per year, serves four objectives. It reduces the military threat from the Russian submarines; and it reduces the linked environmental threat. It provides employment to a skilled Russian workforce, which might otherwise look for less desirable activity elsewhere. It also helps to maintain a shipyard capability that is now being used on a civil commercial basis, servicing Russian fishing vessels and vessels from the neighboring Murmansk Atomflot civil icebreaker facility. While this does not yet enhance current Western commercial prospects, the maintenance of a viable shipyard industry in the Kola Peninsula could well do so in the longer term—for example, in servicing a Nordic fishing industry in response to a reduction in the nuclear environmental threat to that industry.

Metals Recycling

We again illustrate this using the Nerpa shipyard example. The submarine decommissioning process releases quantities of uncontaminated and contaminated metals such as titanium, copper, and steel. These can be recycled (decontaminated as necessary) and sold on a commercial basis at open market prices. There are an estimated 250,000 tons of titanium in the Russian submarine fleet (Typhoon class). The commercial margins would be tight, depending on the cost of decontamination and recycling. However, the Nerpa example again helps to keep the workforce employed and to maintain a local civil shipyard capability. Any commercial advantages for Western companies may be complicated by alleged Mafia involvement in the Russian metal recycling sector. BNFL has examined project proposals but has not yet identified a commercially viable way of taking them forward. However, given the high value particularly of titanium, this may be an area with future potential.

Russian Defense Ministry Conversion Program

The Russian Defense Ministry's Committee for Military-Technical Policy runs a program for commercial conversion of Russian excess defense material. Examples in the nuclear sector include minireactors (uranium-plutonium-mononitride-fueled Brest-300 and Brest-1200) and systems for localization, decontamination, and purification of radioactive waste. At this stage the commercial advantages for Western companies are not clear.

Redeployment of Staff from Russian Closed Cities

In the 10 MINATOM closed nuclear cities, there are 125,000 workers, 750,000 residents, and the largest concentrations of weapons production capability and know-how, weapons-usable material, and HEU and plutonium processing capability. In the current Russian economic situation the caretakers of nuclear materials and keepers of nuclear secrets see little prospect of adequate employment to pay for the basic necessities of life. The closed city infrastructure lacks mobility of labor, thus preventing workers from seeking comparable employment elsewhere, as they might do in a Western market economy. In addition to long-term economic advantages of keeping this skilled pool of nuclear expertise employed in internationally desirable activity, it is in the international security interest to help the Russians downsize their weapons complex to be more consonant with the lower levels of strategic arms already achieved.

The purpose of the 1998 U.S.-Russian Nuclear Cities Initiative (NCI) is to facilitate "civilian production that will provide new jobs for workers displaced from enterprises of the nuclear complex," by directing nuclear weapons workers to non-military scientific and commercial activities. MINATOM envisages redeploying 35,000–50,000 workers over 5 to 10 years, inside or outside the existing city perimeters. Projected NCI funds (currently $30 million per year for 5 years) are quite limited. It is therefore critical that the broad objective of downsizing the nuclear weapons complex itself be coordinated with and also supported through other nuclear security and nuclear reduction programs such as materials disposition, HEU purchase and transparency, and MPC&A. NCI funds are being allocated to institutes, city administrations, and independent enterprises, with the focus on creation of commercial projects, for example, in the energy and scientific sectors. The NCI implementation would call on international governmental, multilateral, and business experience and would involve business and management training for community leaders.

However, we believe that where resources are limited, it is important that efforts to keep nuclear scientists and technology from finding undesirable markets should be focused on those people and facilities where the risk is high. We estimate that only a small percentage of the 125,000 workers have knowledge that would be of direct strategic value to nonnuclear weapon states seeking a covert nuclear capability. A component of NCI should be designed to focus on the nuclear scientists and the research capabilities within Russia's closed nuclear cities. These represent the greatest embodiment of nuclear secrets and therefore a great people risk, yet these are the very people least likely to be immediately successful in direct commercialization.

The national laboratories as well as the Russian-American Nuclear Security Advisory Council (RANSAC) have been looking at ways of facilitating conversion activities at five key Russian nuclear facilities. These include the two nuclear weapon design laboratories, Arzamas-16 and Chelyabinsk-70; and the three plutonium production and separation facilities, Tomsk-7, Krasnoyarsk-26, and

Chelyabinsk-65 (Mayak). Areas of cooperation being examined include nuclear security in the disarmament program, product commercialization, and environmental restoration. For example, Russian scientific experience of nuclear defense pollution could be redeployed on groundwater dispersal systems. This could have longer-term commercial advantages, as in the examples in the Russian Northwest cited above. However, as noted above, conversion of such large numbers of defense workers to commercial ventures will not happen quickly. A complementary component of NCI should be for governments and industry to contract directly with the institutes in the closed cities for research programs in the above areas. At the same time, participating governments would expect to see evidence of downsizing the classified facilities in these cities.

There is also a need to redeploy skills in the Russian civil nuclear research establishments because these also contain sensitive nuclear material and know-how. Here it is marginally easier to deploy these skills for commercial purposes because there is already civil expertise. As an example, BNFL has outplaced civil R&D in Russian nuclear research establishments, with the involvement of the International Science and Technology Center (ISTC). This work has been of the order of $1 million per year and has been done on a commercial basis (i.e., skilled research work at a competitive cost), and there are almost certainly greater opportunities for contract research to be undertaken by Russian scientists.

It is clear, however, that there are no quick and easy answers to the tasks of commercializing the infrastructure and employing the large number of workers in the closed cities in current Russian economic conditions. The path to commercialization will be long, hard, and slow.

Summary

Since the end of the Cold War there has been a massive excess nuclear defense infrastructure in Russia. This excess manifests itself in the form of nuclear weapons-grade material, submarines, spent fuel, radioactive waste, shipyard capacity, and large numbers of skilled staff in secret defense establishments. These collectively pose major security, environmental, and socioeconomic stability threats. In its present state, the developing Russian market economy, with limited mobility of labor, cannot pay the tens or hundreds of billions of dollars for the necessary remedial action. There is a clear Western imperative to help, in close partnership with the Russian authorities. Except where there are significant sources of external funding, there are few easy commercial paths for Western or Russian companies, and the defense conversion and commercialization program will take years to achieve.

However, the process of conversion and environmental restoration can create a new climate for commercial activity, including countertrade, especially in the regions that are now beginning to cooperate economically with their neighbors. The real key is to attract foreign investment to help transform some of the defense infrastructure to supply international nuclear services. The future health and consolidation of the Western nuclear industry and the commitment of the Russian government to introduce radical reform in Russia's laws and regulations in order to seize these opportunities will be decisive factors.

Recommendations

Western governments must continue to focus on three strategic objectives: achieving safe and secure materials management, avoiding the transfer of nuclear weapons know-how and technology to undesirable international markets, and addressing potential or actual environmental hazards of international significance.

We believe, however, that international leadership and coordination are lacking at the current time and strongly recommend that Western governments give this urgent attention.

Governments must also lay the legal and regulatory foundations to allow the Western nuclear industry to explore longer-term commercial opportunities in tackling the problems of the Russian excess nuclear defense infrastructure. Specific recommended actions are for:

- Russia to relax relevant domestic barriers and attitudes to closed cities to allow reliable, timely, and sustained access by foreign commercial companies that seek to support Russian commercialization of the defense infrastructure. With respect to Russian legitimate security concerns, relevant barriers have to come down in order to facilitate the necessary commercialization.
- Russia to accelerate the pace of change in its tax, customs, and liability laws to provide a climate for investment and enterprise. Foreign companies will not invest in Russia until this happens.
- Russia to establish an integrated defense conversion plan, including a nuclear materials stewardship program, with the objective of separating its defense and civil activities and reducing the number of sites. Until there is complete transparency and international verification of its civil (peaceful) activities, Russia will not be able to fully exploit international commercial opportunities in the nuclear services market. We believe that foreign investment is the key to the commercialization of the excess defense infrastructure, but it is unrealistic to think that the entire Russian nuclear complex can be transformed in this way. Russian and international resources must be targeted to have any chance of success and some facilities will have to close.
- Western governments to facilitate the commercialization process by supporting Russian initiatives to win international civil business. Many governments will insist on concluding nuclear trade agreements with the Russian Federation if they or their national companies desire to cooperate with Russia in the various aspects of the nuclear fuel cycle, including spent fuel storage or waste disposal. A peaceful nuclear cooperation agreement with the United States is a prerequisite for the importation into Russia of U.S.-origin special fissionable materials for processing or storage and of certain types of equipment. This will require the two governments to reach a solution to their differences over Russian

assistance to the Iranian nuclear program. Failure to achieve this will encourage Russia to seek markets where the United States has disengaged or has no political control.

- Western governments to remove regulatory and other obstacles to the international consolidation of the Western nuclear industry, which needs government support and encouragement in pursuit of these long-term political and security objectives. A significant component of the Russian commercialization process will involve multinational companies of international standing, which can act in partnership with the Russian government and verification agencies to provide international assurances over product quality and peaceful use.
- All governments to recognize the important environmental role that nuclear power already plays in mitigating climate change (caused by the burning of fossil fuels) and promote the expanded use of nuclear power where it is safe and economic. The building of advanced nuclear power stations (having the ability to consume excess weapons material) will alleviate international concerns about the oversupply at the front end of the nuclear fuel cycle and increase the market for down-blended excess Russian HEU without damaging other Western suppliers. Consideration should be given to forging alliances with Western reactor suppliers to ensure that new generation plants also have the requisite safety features. We believe that the recent contracts for the purchase of Russian HEU have been a marked success and that additional sales of down-blended LEU could be increased under the proper circumstances with the profits being used to finance the disposition of Russian excess weapons plutonium.
- Russia and the United States to reexamine their plutonium disposition programs to establish the scope for integrated, bilateral cooperation. There seems no reason why it should be necessary to duplicate all plutonium-related disposition facilities in both Russia and the United States. Such an integrated program could save very significant amounts of money if there is the political will to do so. This and other bilateral initiatives should establish an international basis for the fungibility of nuclear materials based on factors such as fissile content and radioactivity in order to avoid the needless transport of nuclear materials.

Overall, we recognize the difficulty in achieving these objectives; commercialization will be long, hard, and slow. But there are real opportunities, given strong political leadership and a will to make progress. We believe that world peace and security deserve that effort.

CHAPTER 4

Nuclear Materials Transparency

Task Force IV

THIS CHAPTER SERVES AS AN INTRODUCTION TO THE TOPIC OF TRANSPARENCY IN SUPPORT OF THE VISION OF GLOBAL NUCLEAR MATERIALS MANAGEMENT.

Transparency, as openness, is not a new concept. The development of transparency measures designed to promote openness and confidence has long been held as desirable. Although pursued in support of arms control during the Cold War and modestly furthered in arms accords relating to such provisions as data exchanges and noninterference with national technical means (NTM), openness did not make major breakthroughs in the climate of secrecy in the Soviet Union until the mid-1980s. Since that time, amid the political changes in the FSU, the scope and prospects of transparency have broadened.

In recent years, the interest in and debate over transparency go beyond classical arms control and appear to be moving toward global issues, in part because of growing concerns about the security implications of the proliferation of nuclear, biological, and chemical weaponry. Concerns about nuclear proliferation have focused new attention on nuclear materials, in particular materials made excess to defense needs by the changes in the international security environment. The end of the Cold War has resulted in substantial changes to the way the nuclear powers view their stockpiles of weapons and weapon materials, and they are beginning to make changes in the directions of their programs.

There seems to be a growing willingness on the part of most of the established nuclear powers to reduce the sizes of their respective stockpiles and to use the excess material in peaceful energy production or to provide for its ultimate disposal. In this context, the U.S.-Russian Helsinki Summit language reflects a growing sense that warheads and materials need to be better controlled. Apparently this view has arisen out of the requirements of the CTR program, laboratory-to-laboratory programs on material protection control and accounting, and other activities between the United States and Russia to enhance the security of the warheads and materials from the old Soviet arsenal. Also the START process will likely lead to requirements related to warheads and materials as numbers of delivery systems are reduced dramatically.

Previous arms control transparency measures and IAEA approaches to safeguards can provide valuable precedents in addressing current opportunities.

With this changing global environment and unprecedented responses to its challenges, the international nuclear community needs to adopt a collection of methods to face the future in a positive way while it addresses negative developments like the South Asian nuclear tests. Transparency may be a key tool for addressing new issues; it is certain to become more significant, both as a result of and outside of the arms control and nonproliferation processes. For the Russian nuclear program in particular, the concept of transparency may provide assurances to others that all that can be done is being done in providing safety and security to nuclear materials, facilities, and expertise.

Transparency: Concept and Issues[1]

A universal understanding of the meaning of transparency does not exist even within the arms control and nonproliferation communities. For the purposes of this report, the following definition of transparency is assumed:

Transparency is a cooperative process that is based on thorough risk–benefit assessments and that (1) increases openness and builds confidence; (2) promotes mutual trust and working relationships among countries, national and international agencies, and the public; and (3) facilitates verification and monitoring measures by information exchanges. These exchanges would be related to safety, security, and legitimate use of nuclear materials consistent with protection of national security and proprietary interests.

As the definition suggests, transparency is more than a picture of a nuclear program or a specific site. It permits the accumulation of data, both direct and indirect, over an extensive period of time to build confidence that behavior of a country or collection of countries is consistent with the agreements or norms.

Transparency versus Opacity

Transparency needs to be viewed in the context of relationships between nations where, with few exceptions, each nation tends to be somewhat opaque, or secretive, in how it conducts its business, especially in nuclear matters. There are often reasons for this apparent opacity that go beyond national security reasons. Differing cultures, legal constraints, geography, protection of proprietary business data, and differing norms for freedoms of the press are all factors that produce a natural opaqueness of a society, especially when viewed by another. Only when certain conditions are met will a country allow the thinning or elimination of the protective opaqueness.

A country's opacity to others has been, and can be in the future, an impediment to further progress in arms control and nonproliferation regimes. Increasing trans-

1. The following material is derived from Joseph F. Pilat, "Transparency and the New Verification Agenda," in *Arms Control Issues for the Twenty-First Century,* ed. James Brown, SAND97-2619 (Albuquerque, N.Mex.: Sandia National Laboratories, 1997).

parency can be a difficult and slow process. Each nation must decide how agreements impinge upon its sovereignty and how it will manage the risks to its security associated with increased transparency. In the process of reducing opacity through greater transparency, the emphasis must be on cooperation and collaboration among states that are seen as equal partners in this endeavor.

Transparency, Monitoring, and Verification

Transparency is not identical to monitoring or verification. Monitoring is the technical process of gathering the data allowed under any agreement or regime. That process can include everything from inspectors physically on-site at a declared facility, to actual readings from sensors placed to observe activity, to the analysis of material samples. Verification is a political process that involves authoritative judgments about compliance, with commitments based on the data and interpretations provided by the monitoring community. Transparency, or the permitted knowledge of areas and information otherwise opaque, can be of enormous importance in increasing the effectiveness of both monitoring and verification.

Trends, Risks, and Limits of Transparency

Transparency is being furthered by political and technological trends. Political developments such as democratization are opening once-closed societies. Technological developments, including the information revolution and the surge in commercial surveillance technologies, are also creating a new openness. These trends, political and technological, are coalescing to produce a world in which transparency is increasing—irrespective of any transparency initiatives, negotiations, or agreements.

Still, transparency has risks and limits, which are not always given their due in the literature. Among the risks are the great asymmetries in transparency around the world. The open societies of the West are very different from both dictatorships and developing countries. These asymmetries caution against an uncritical hope that there is no longer a need to pursue transparency as a matter of policy. Other risks include the prospect that classified, sensitive, or proprietary information will be compromised or released (with an adverse impact on national security and international obligations); the possibility of the information channels being used for misinformation from the other party or parties; the creation of a false sense of confidence; the questionable value of information obtained compared with intrusive verification; and the like. Nonetheless, opportunities to increase transparency do exist, and any benefits offered must be balanced with the risks. The components of this balance sheet are addressed below.

GNMM: A Transparency Regime?[2]

Global nuclear materials management (GNMM) is a vision of the effective management of civilian, defense, and excess defense nuclear materials worldwide to ensure safe, secure, and transparent use of these materials from cradle to grave.[3] The near-term goals are to

- continue to build on the excellent U.S. record of success in the safety and security of civilian materials;
- incorporate excess defense materials into the procedures and processes for materials management; and
- expand the perspective of organizations owning materials or involved with materials management to include transparency in an efficient manner.

If these activities are fully realized, the resulting environment—the product and goal of GNMM—will support the future of nuclear energy, nuclear arms reductions, and nuclear nonproliferation.

Seen from this perspective, GNMM is not exclusively a transparency regime, although transparency is a key feature that will be highlighted below. GNMM has safety, security, and verification elements, all of which can be served by well-designed transparency measures. In this context, under an effective GNMM, the international community will be charged with supporting transparency measures to provide confidence to all appropriate parties that at all times the handling of nuclear materials meets global norms for safety, security, and assurance of declared use. This will apply to production, storage, processing, transportation, and disposition of these materials.

Although GNMM remains a vision, many cooperative steps have already been taken to improve and ensure the effective management of nuclear materials. The IAEA has long served the community as a resource for technical advice and state-of-the-art nuclear material safety, control, and accounting practices. The agency is expanding the role of transparency in its safeguards and other missions. Still, it remains true that in many cases the international conventions and agreements guiding materials management are voluntary and contain no verification or transparency measures.

The international community's involvement in development of effective standards for safety, security, and transparency of all materials should increase, and the resulting standards should become an international norm, with appropriate regional measures designed to address unique problems and issues. The IAEA has recently developed a set of standards for safety and security for civilian nuclear

2. Steve Dupree et al., "A Uniform Framework of Global Nuclear Materials Management" (paper presented at the 1999 annual meeting of the European Safeguard Research and Development Association, Saville, Spain).

3. International concern about materials management traditionally has been limited to civilian materials. Defense materials and excess defense materials are now also being considered in this context. At present, this new material is in transition: defense material is being declared excess, and excess material is being managed through a transition to civilian use or by other means.

material that could serve as a benchmark. National, regional, and global strategies for disposition of weapons legacy materials will mature, and a regime that evolved to ensure transparency in that disposition is desirable. All states that possess nuclear materials, whatever their policy toward nuclear power and nuclear weapons, should help build and subscribe to an effective international transparency regime with appropriate inspection and monitoring rights.

As noted above, this could in practice be supplemented by regional approaches. For example, a regional nuclear energy cooperation regime in Asia, as embodied in proposals for an ASIATOM (an Asian atomic energy community) or PACATOM (a Pacific Rim atomic energy community), may eventually be agreed to and could benefit from the example and experience of European Atomic Energy Community (EURATOM). Although the declared objectives of these and other regional nuclear proposals are not always clear, many seem designed primarily to promote transparency in nuclear-fuel cycles, particularly those that involve direct-use nuclear weapon materials.

Achieving the vision of GNMM will require additional effort and cooperation among all parties involved in materials management. If this is forthcoming, GNMM could become not just another technique for materials monitoring, but rather a new, cooperative way of doing business.

Transparency's Risks and Benefits

If transparency is to play the desired role in GNMM, it is essential to fully understand the risks and benefits of these measures. A thorough analysis of specific transparency-measure proposals will ultimately be necessary. In the following, general benefits or incentives for transparency are discussed, followed by potential risks, disincentives, or impediments to transparency measures.

The benefits or incentives for greater transparency are reasonably clear. Transparency can provide confidence that a state is behaving in a certain fashion or that its activities are in conformance with certain agreements, standards, and norms. In the nuclear materials arena, for example, transparency measures provide information to outside parties to facilitate independent assessments of the safety, security, and declared use of nuclear materials, and to ensure that all appropriate measures have been applied to provide for assurance and confidence both domestically and internationally that the material is being used as declared and that all safety and security measures are in place. In this case and others, transparency can offer these benefits, in principle, at a relatively low cost and in political circumstances where other means of reducing suspicions (e.g., verifiable arms accords) are not politically feasible. Trends in technologies and, to a large extent, politics have opened the way for ever-increasing transparency and make specific targeted measures for GNMM or other objectives more obtainable and realizable.

The risks of impediments to transparency are less well understood, but understanding them is vital to analyses of the desirability of general and specific transparency measures. First, it must be recognized that different states may have very different levels of openness owing to cultural, economic, legal, and political factors. This asymmetry may make it difficult to develop effective, mutually beneficial transparency measures. In practice, it limits the prospects of transparency, especially for the countries of the world that are least open.

National security considerations, including the protection of classified or sensitive information, are important impediments to transparency measures and point to real security risks. The misuse of transparency to foster disinformation through incomplete or false information or to improve, say, the targeting of an adversary based on exchanged information is a great concern that must be addressed. The prospect of proliferation-relevant information being compromised by transparency measures is another problem, one in which national security and international agreements may both argue against certain types of measures. In addition to these concrete national security concerns, the prospect of transparency providing a false sense of security should not be dismissed.

In similar fashion, operational security (OpSec) and safety constraints also pose limits of openness and transparency. With respect to OpSec concerns, there is often a conflict between the desire for more transparency and the need for security against terrorism. The transport of nuclear materials both domestically and internationally provides excellent case studies. In Germany, demonstrators damaged railway tracks and roadways along which irradiated fuel travels to depositories, and the shipment of Japanese MOX fuel through the sea lanes required the balancing of routing information releases and the security of the shipping vessel.

The concern that transparency measures will compromise proprietary information, exports, and other commercial interests has been a significant impediment to transparency, as evidenced, for example, by IAEA safeguards. In the future, transparency measures will have to be balanced against these concerns.

Although technological developments have largely been an incentive to transparency, some technologies, including those involving control and accountancy of nuclear materials, have inherent opacities. The properties of these technologies include their statistical nature (measurement uncertainties), the effects of which may accumulate over time. The uncertainties associated with technological measurements can pose problems for transparency. These uncertainties can often be calculated, but in some cases they are so significant that they may prevent or reduce the value of proposed transparency measures or create mistrust even among the very constituencies the technologies are designed to assist.

Policy Recommendations

In order for the concept of GNMM to be furthered, a series of transparency measures of differing impacts and levels of difficulty is desirable. This section identifies a number of potential transparency measures or activities that could move current nuclear material efforts toward the vision of GNMM. The categorization in terms of near-term (6–12 months), mid-term (1–3 years), and long-term (3–5 years) refers primarily to implementation timing although there is a correlation between impact and level of difficulty in achieving the proposed measure/activity. In addition, each recommendation has been prioritized within each time element so that the policy community can understand the relation among recommendations.

Near-Term Recommendations

- **Make the U.S.-Russian relationship a model for others to emulate.**

  The United States and the Russian Federation possess the world's largest nuclear arsenals and nuclear material stockpiles. Although there have been significant steps taken since the end of the Cold War to reduce those arsenals and secure the fissile material declared excess to defense needs, the arms control process is currently in a state of flux owing to a host of other political and economic factors. Bilateral initiatives to obtain these objectives have also been affected by the instability in relations. However, the United States and Russia should not delay in putting in place a set of tools and measures that will enhance transparency activities and help shape the U.S.-Russian relationship into a model for others to emulate.

  Specific transparency elements directly related to this bilateral association are addressed below. These bilateral transparency elements focus on four specific areas: the incorporation of transparency activities into the U.S.-Russian Nuclear Warhead Safety and Security Exchange agreement (WSSX), the revitalization of information exchanges on nuclear stockpiles, transparency experiments at nonsensitive facilities, and the development of transparency norms and standards for the safety and security of excess and defense nuclear materials. The successful completion of these near-term elements will require extensive cooperation between the United States and Russia, but if successful, will go far to setting the stage for more ambitious transparency recommendations.

  Element One: Renew/renegotiate the WSSX to include the appropriate provisions for the continuation of the Russian-U.S. laboratory-to-laboratory transparency activities.

  GNMM coverage: defense and excess nuclear materials

  Priority: critical

  U.S.-Russian laboratory-to-laboratory transparency initiatives, either being considered or under development, are virtually stalled owing to Russian concerns over security and the lack of a government-to-government agreement that addresses transparency. In addition, the WSSX agreement will expire in 1999 if not renewed.

  WSSX should be extended and expanded to include the transparency work being currently accomplished under the laboratory-to-laboratory program. With the inclusion of transparency as part of WSSX, the U.S. and Russian technical experts will have the governmental authority to reenergize their work. Using this authority, they will develop and evaluate mutually acceptable measures and procedures for transparency of strategic nuclear warheads and their elimination, as committed to by Presidents Clinton and Yeltsin in Helsinki in March 1997 and in Cologne in June 1999.

Element Two: Implement U.S.-Russian agreements on data exchanges on the aggregate stockpiles of nuclear warheads, on stocks of fissile material, and on their safety and security.

GNMM coverage: defense nuclear materials

Priority: urgent

Numerous U.S.-Russian summit agreements have called for data exchanges as noted above (Moscow, May 1995; Washington, September 1994); however, the implementation has never been completed. Renewed efforts to complete such exchanges are needed and will provide a solid starting point for further cooperation in transparency. The exchange of this information could over time enhance national and international security by improving both communities' understanding of how the Russians conduct their nuclear business. Although the United States has made numerous unilateral declarations in the past, we would see these data exchanges going beyond those one-time events in scope and frequency as appropriate. Therefore, these exchanges and other appropriate transparency measures could also enhance Russian understanding of the U.S. nuclear stockpile. Transparency measures to serve as a check on the accuracy of the official declarations could initially include limited technical exchanges, site visits, and tours with a view to expanding coverage of declared facilities and sites as the relationship grows.

Element Three: Support a series of joint bilateral U.S.-Russian transparency experiments, starting at nonsensitive facilities and with nonsensitive activities.

GNMM coverage: defense, excess, and civilian nuclear materials

Priority: urgent

Joint U.S.-Russian transparency experiments could include areas such as technology-driven transparency measures, monitoring procedures, safety and security, risk–benefit analysis, information management and delineation of procedures, and processes for access management. These potential experiments would serve as a foundation for development of international transparency norms. Participation of Russian weapon scientists in these joint experiments could reduce some of the concerns over these experts selling their expertise elsewhere. However, this activity must be sensitive to classification and to the risk of measures being rejected outright by the Russians in light of current U.S.-Russia relations.

Element Four: Promote the development of bilateral transparency norms and standards for the safety and security of excess and defense nuclear materials.

GNMM coverage: defense and excess nuclear materials

Priority: urgent

Utilizing the data developed in the joint experiments noted above, the United States and Russia should hold a series of workshops designed to develop transparency norms and standards for safety and security of excess and defense nuclear material and associated transparency norms for each area. The work in the MPC&A program, as well as existing related IAEA standards, may serve as starting guidelines. The development of these norms and standards could lay the groundwork for eventual implementation and for future examination for broader application to other nuclear states.

- **Complete U.S.-Russian-IAEA trilateral agreement.**

  GNMM coverage: excess nuclear material

  Priority: critical

  Completion and early implementation of the agreement will enhance irreversibility of excess declared nuclear material through transparency and provide some level of verification through an international agency. This agreement will provide a valuable step in assuring the international community that indeed nuclear materials are not being diverted back into the weapons programs of Russia (or the United States). In addition, if this agreement is successful, it could serve as a model for any future efforts to include the British, French, or Chinese nuclear programs regarding the transparent handling of any nuclear materials they have declared, or may in the future declare, to be excess to their weapon programs.

- **Increase support for expanding IAEA transparency efforts, especially in strengthening safeguards, integrating safeguards, and safety and physical protection.**

  GNMM coverage: civilian nuclear materials

  Priority: important

  The United States should commit sufficient technical expertise and resources to assist in timely and effective implementation of IAEA activities that increase transparency in the areas of safeguards, safety, and physical protection. The new IAEA-strengthened safeguards system and the effort to integrate safeguards measures, including those available under existing authority and those available under the additional authority of INFCIRC/540,[4] are formidable challenges for the IAEA, especially when one considers that the objectives of this integration are to improve the effectiveness of safeguards, increase the efficiency, and reduce costs. Yet these steps will be essential for eventual transparency of IAEA safeguards. Evaluating design concepts for an integrated system, developing and applying technologies that will ensure the effectiveness of integrated safeguards, and developing models for proper risk assessments are examples of key areas where the IAEA and the international community would benefit from increased U.S. assistance.

---

4. "Model Protocol Additional to the Agreement(s) Between States and the IAEA for Application of Safeguards," Information Circular 540 (Vienna, Austria: IAEA, December 1998).

Mid-Term Recommendations

- **Place all nuclear materials declared excess by the P-5 under U.S.-Russian-IAEA trilateral-type agreements.**

  GNMM coverage: excess nuclear materials

  Priority: critical

  This action by the permanent members of the UN Security Council [P-5]—United States, Russia, China, France, and United Kingdom—would capture material declared excess to their military needs, ensure the irreversibility of excess materials through transparency, and provide some level of verification through an international agency. This agreement will provide a valuable step in assuring the international community that, indeed, nuclear materials are not diverted back into the weapons programs of these states.

- **Implement U.S.-Russian safety, security, and transparency norms and standards for excess and defense nuclear material developed in the previous activities.**

  GNMM coverage: defense and excess nuclear materials

  Priority: urgent

  Implementation of the bilateral norms and standards would be a watershed event in the arena of transparency. For the first time both the United States and Russia would have viable tools to make assessments about the status of each other's nuclear arsenals and how they are being protected. As an example, these tools should go a long way toward gaining confidence that unauthorized use or launch has been minimized. Depending on how the norms and standards are publicized, there could also be greater assurances to other nations that the United States and Russia are proper stewards of their nuclear weapons. Moreover, they could be a model for broader regional and even international standards.

- **Improve with cost-effective technologies the international safety, security, and transparency standards for nuclear materials in spent fuel and waste from P-5 civilian and former weapon-material-producing fuel cycles.**

  GNMM coverage: civilian and excess nuclear materials

  Priority: urgent

  This intermediate step addresses complex, difficult problems that are at the center of the public debate and concern over nuclear issues, and can thereby contribute to nuclear energy development, nonproliferation, and nuclear arms reductions. The commitment by the P-5 to ensure that spent fuel and waste (from civilian and former defense activities) are addressed in accordance with international standards promoting transparency would also reduce fears about safety, environmental, and proliferation dangers. It may further mitigate perceptions of discrimination among Nuclear Non-Proliferation Treaty (NPT) nonnuclear-weapons states.

- **Negotiate, conclude, and implement a data exchange among the P-5 regarding nuclear weapons inventories and support of infrastructure modeled after the U.S.-Russian efforts.**

  GNMM coverage: defense and excess nuclear materials

  Priority: important

  In the form of well-managed tours and inspections, coupled with official declarations, the transparency of these countries' nuclear weapons inventories and supporting infrastructure could enhance overall strategic stability and possibly provide a demonstration that these countries are living up to their NPT commitments. Obviously, not all data could or would be shared owing to national security issues; however, sufficient data should be available to highlight that each country is sustaining the proper stewardship of nuclear weapons under its control.

- **Promote among nuclear weapons states the development of transparency norms and standards for the safety and security of defense nuclear materials.**

  GNMM coverage: defense nuclear materials

  Priority: important

  The British, French, and potentially the Chinese should hold a series of workshops designed to develop norms and standards for the safety and security of their defense nuclear materials and associated transparency norms for each area. Although there are considerable differences between the U.S. and Russian nuclear programs, the model developed by these countries could serve as a benchmark. The development of such norms and standards would contribute to the fact that each of these nations is managing its stockpile of warheads and nuclear materials in a responsive manner.

- **Reexamine the current U.S. linkage, in START III policy statements, of the numerical reductions of strategic nuclear forces and the irreversibility of force reduction through transparency of strategic nuclear warhead inventories and the destruction of strategic nuclear warheads.**

  GNMM coverage: defense and excess nuclear materials

  Priority: important

  The linking of strategic force reductions and ensuring that the warheads reduced or eliminated with those force reductions are not reintroduced into the defense posture as noted in the Helsinki Summit and Cologne Summit statements was a sound strategy when developed. However, if the START II ratification process remains stalemated, it would be useful to revisit this linkage issue. If, in fact, one could make some progress in declarations as noted in the near-term recommendations, element two, then more extensive transparency measures might be possible regarding nuclear support infrastructure and warhead dismantlement if the linkage is decoupled. Some part of the transparency measures could be based on technology efforts being considered or under development in the U.S.-Russian bilateral initiatives.

Long-Term Recommendations

- **Establish international safety, security, and transparency norms for all states' nuclear fuel cycles.**

The United States and Russia are reducing the sizes of their nuclear weapons stockpiles, resulting in a growing level of direct-use material that is excess to national defense needs. Concurrently, around the world, the amount of material at the back end of nuclear fuel cycles is growing at an ever-increasing rate. Other countries are developing and implementing fuel cycles that will profoundly increase the amount of nuclear material. The result of these trends is a growing level of materials that will outstrip current political, regulatory, and technical monitoring capabilities. It is therefore necessary to develop an international consensus on proper management standards for all nuclear materials.

Element One: Negotiate and initiate safety, security, and transparency norms and standards for defense materials within the P-5 community.

GNMM coverage: defense nuclear materials

Priority: important

Through this activity, the P-5 countries will exchange safety and security approaches for defense and excess materials. Equivalent approaches will be agreed upon that will allow each country the flexibility it needs to meet jointly developed standards. Participation in this activity will be restricted to P-5 member countries because of the sensitive nature of the materials and processes involved. The P-5 will develop a mechanism for communicating with international agencies and individual countries to raise confidence levels worldwide.

Element Two: Negotiate and initiate international safety, security, and transparency standards for materials in the civilian fuel cycle.

GNMM coverage: civilian nuclear materials

Priority: important

Through this activity, those nations with civilian nuclear industries in cooperation with the IAEA will jointly develop common transparency standards for safety, security, and legitimate use of materials and processes. This process will utilize current and developing IAEA standards and norms as a benchmark, with a view to furthering them. Here the United States can lead by example. To encourage others to negotiate and implement transparency standards for safety, security, and legitimate use of materials with the IAEA, the United States could serve as the test bed for developing and testing those standards at one or more U.S. facilities. The United States and IAEA could conduct experimental exercises and testing for sampling regimes, examine different access level requirements, analyze the associated data, and develop cost impacts for implementing the standards at civilian facilities. The experience the United States would gain working these types of standards with Russia and the other P-5 members would be valuable to the IAEA in this vital effort.

In addition to improving international standards and norms in these areas, regional approaches should also be examined. Recognizing that each region may need to address unique issues and may require additional (and more intrusive) measures, the international standards will be broad and will serve as a foundation upon which regional agreements are built.

Conclusion

Changes in the world have affected the way we look at nuclear materials and the management and control regimes that surround them. The nuclear enterprise is evolving slowly toward greater openness and transparency in all the areas of interest—defense nuclear materials, excess nuclear materials, and civilian nuclear materials. The vision of GNMM is to provide greater transparency regarding safety, security, and legitimate use for the entire spectrum of nuclear materials, from cradle to grave. Nonetheless, there will remain significant limits on transparency for at least defense materials for the foreseeable future.

Transparency is a key element in that vision, and there are several steps that can be made in the near term to implement steps toward GNMM. Transparency offers no panacea, however. It can supplement verification measures, but to the extent that transparency is put forward as a substitute for verification, it is important to understand that this approach will not offer all of what we have come to expect from verification during recent decades. In similar fashion, transparency of safety and security will be welcome, but it is not the same as mandated standards. As a result of such considerations, it would seem that transparency would not, for the foreseeable future, be all-encompassing in the GNMM context. Nonetheless, openness and transparency measures have in recent years been widely perceived as increasingly germane to addressing emerging materials issues. They will undoubtedly become more significant in this regard over time.

CHAPTER 5

U.S. Domestic Infrastructure and the Emerging Nuclear Era

Task Force V

SINCE THE END OF WORLD WAR II, THE UNITED STATES HAS TAKEN THE LEAD IN ESTABLISHING INTERNATIONAL TREATIES AND INSTITUTIONS TO MINIMIZE THE SPREAD OF NUCLEAR WEAPONS, FOSTER THE USE OF PEACEFUL NUCLEAR TECHNOLOGY, AND MONITOR AND CONTROL FISSILE MATERIALS.

With the end of the Cold War, the United States has entered into agreements with Russia to reduce nuclear armaments, with a view toward ending the nuclear arms race. These achievements have not been perfect, in part because the balance between nuclear arms control and peaceful nuclear technology has not been consistently maintained. This consistency should be restored and greater focus placed on the continuing work necessary to realize a stable GNMM regime in this second nuclear era.

It is unfortunate that at this crossroad between the first and second nuclear eras the United States is withdrawing from effective international leadership and its nuclear technology infrastructure is fading. In light of this situation, the following recommendations are proposed to strengthen U.S. leadership and improve the implementation of largely bipartisan international nuclear policies in fissile materials control and nuclear arms reduction, international commercial nuclear plant safety, and spent fuel and radioactive waste management.

- Accelerate excess weapons materials disposition;
- Strengthen support of IAEA safeguards responsibilities and its additional protocol, by expanding the scope of the Convention on Physical Protection of Nuclear Material and its supporting guidelines[1] to cover storage facilities within national boundaries;

1. "The Physical Protection of Nuclear Materials and Nuclear Facilities," INFCIRC/225/Rev. 4 (corrected) (Vienna, Austria: IAEA, June 1999).

- Develop an international consensus on proliferation resistance standards in commercial nuclear power;
- Increase G-7 funding and coordination of former Soviet bloc reactor safety;
- Support the IAEA reactor safety convention;
- Promote regional storage facilities; and
- Support the IAEA joint conventions on the safety of spent fuel and radioactive waste management.

Of key importance are additional recommendations that domestic nuclear energy policy be redirected with the overall objective to "...create a broad-based nuclear mission that advocates a viable commercial sector, produces top-quality scientists and engineers, develops options and techniques for disposal and recycle, and funds reactor and fuel cycle R&D as an integrated component of an overall energy portfolio," as proposed by the Decision Makers' Forum sponsored by the U.S. Senate Nuclear Issues Caucus on a New Paradigm for Nuclear Energy.[2] This will require a series of timely domestic actions:

- Establish uniformity in safety and environmental regulation among all power generation alternatives;
- Eliminate overlapping regulatory jurisdictions;
- Improve the effectiveness of nuclear power regulation by moving to performance-based, risk-informed regulation and processing license renewals expeditiously;
- Improve national spent fuel management by providing a centralized interim storage facility for retrievable spent fuel and accelerating the Yucca Mountain Site Viability Assessment follow-up work;
- Increase low-level radioactive waste storage capability; and
- Realistically define low-level radiation health effects.

A redirection of policy is also needed for the long-term realization of a dramatically expanded supply of nuclear fuel by transmutation of the fertile isotopes in uranium and thorium that would ensure many centuries of combustion-free power generation. This will require international cooperative reengagement by the United States in R&D on high fuel conversion systems and in establishing international standards and mechanisms to ensure that diversion to nuclear weapons use will not result from commercial deployment.

Two overarching steps will greatly help to achieve these implementation improvements and international policy redirections: a demonstration of international and national government leadership as well as national cooperation among government, industry, academia, and the environmental community. The implied

2. "Report of Decision Makers' Forum on a New Paradigm for Nuclear Energy," INEEL/EXT-98-00915 (Idaho Falls, Idaho: Idaho National Engineering and Environmental Laboratory, June 19, 1998).

changes in mind-set are a small price to pay to realize a stable international regime that will move the world away from nuclear weapons and provide the benefits of nuclear energy.

History of U.S. Global Leadership in Nuclear Energy: Lessons Learned

The United States assumed strong leadership in nuclear energy in the years immediately following World War II, making many proposals to establish international controls to stem the spread of nuclear weapons, but none was acted upon. By the early 1950s, the fact that the Soviet Union and a few other countries had nuclear weapons capability led to the atoms-for-peace proposal by the United States; its goal was to drop the veil of secrecy and substitute transparency of the technology for peaceful purposes in return for pledges by the recipients of the technology that they would not develop nuclear weapons.

A near-term outcome of that initiative and continuing U.S. leadership was the formation of the IAEA as an agency of the United Nations to exercise controls over the spread of weapons and to monitor the technology transfer.[3] After lengthy negotiations, the NPT was approved in 1968. The treaty codified the requirement that the nonnuclear weapons signatory countries would not develop weapons as a quid pro quo for the transfer of peaceful use technology. The treaty also included a pledge by the nuclear weapons powers to end the nuclear arms race.

Technical, Commercial, and Safety and Licensing Data Exchange

The United States implemented the technology transfer agreement, rapidly providing other countries with R&D for peaceful uses. The United States also launched its own nuclear power demonstration program, evaluating a variety of combinations of coolants, moderators, fuels, and pressure-containing concepts.[4] Both the industry and the national laboratories participated, and the programs were jointly funded by industry and government. The U.S. demonstration program narrowed the field for near-term deployment to three reactor concepts (all requiring enriched fuel): light-water-cooled and -moderated (both pressurized water and boiling water), high-temperature-gas-cooled and graphite-moderated, and fast-spectrum-sodium-cooled. The national nuclear power research, development, and demonstration (RD&D) programs of many countries utilized the screening process of the U.S. demonstration program and focused on one or more of these concepts.

This screening process, combined with the rapid development of light–water reactor (LWR) technology through the U.S. effort to develop nuclear powered submarines, motivated U.S. industry to focus on LWRs for commercial electricity production. U.S. reactor manufacturers and architect engineers entered into com-

---

3. D. Fischer, *History of the International Atomic Energy Agency: The First Forty Years* (Vienna, Austria: IAEA, September 1977).

4. John J. Taylor, Chapter 11, in *Nuclear Power: Policy and Prospects,* ed. P.M.S. Jones (New York: Wiley, 1987).

mercial alliances in Western Europe and Japan in the 1960s to build small LWR demonstration nuclear plants.

These alliances were not simply contracts to build the plants but entailed licensing agreements that governed the transfer of technology to the host country and defined the role of the industry in the host country in implementing the project. In effect, they were bilateral international cooperative RD&D programs, with U.S. industry in the leading role. The contracts stipulated that the design be in conformance with the U.S. Nuclear Regulatory Commission (NRC) regulations, providing assurance that a safe plant by U.S. standards would be provided. This assurance was important because there was only a nascent regulatory framework in many of the nations involved. The contract provisions in effect established international recognition and conformance with specific U.S. safety standards. In addition, the U.S. regulations provided a reference base upon which an IAEA international consensus on reactor safety standards could be built and other national regulatory regimes established.

Thus, the combined leadership of the U.S. government and industry established a key international base for the international control of nuclear weapons materials and safety standards for nuclear power plants. In the time since, the IAEA has fostered peaceful nuclear technology transfer to the nonweapons states, has developed international safety standards and codes, and has assumed a vital role in controlling the spread of nuclear weapons.[5]

Success and Failure in Policy-Driven International Cooperation

Much progress has been made in the ensuing 45 years in achieving the aims of U.S. nonproliferation policy. The expansion in the number of weapons states has been much less than predicted. The IAEA has become a significant international force in the control of fissile materials and in the establishment of international nuclear safety and environmental standards. The nuclear arms race ceased with the START agreements and the end of the Cold War. The revelation after the Gulf War that Iraq had evaded IAEA surveillance of their nuclear arms effort led to the institution of the 93+2 program, giving the IAEA stronger surveillance capability, including authority for unannounced inspections.

Yet, the NPT and the IAEA have not been fully successful, particularly in light of events in 1998 in Pakistan and India that threaten the NPT regime. After India's first nuclear bomb test in 1975, the United States fostered a major international study, the International Nuclear Fuel Cycle Evaluation (INFCE), to evaluate proliferation resistance in commercial fuel recycle. Sixty-six nations and five intergovernmental organizations participated in the evaluation.[6] A massive report was completed in 1980, with many proposals for increased proliferation resistance. Yet there has been essentially no cooperative international follow-up on those proposals, nor has any consensus been reached on nonproliferation standards.

5. Ibid.

6. A.T. Grey, "Taking Stock After INFCE," *Uranium and Nuclear Energy* (proceedings of the Fifth International Symposium held by the Uranium Institute, London, September 1980) 5 (1980): 301.

The international deployment of U.S. nuclear plants designed to NRC standards was also marred by the fact that the Soviet bloc did not participate and accordingly developed its own designs and safety standards more quickly than the IAEA could gain consensus on international safety standards. The less rigorous safety basis that evolved for those plants was one of the underlying causes of the Chernobyl accident. This situation has been improved by the gradual assumption of IAEA safety standards by the former Soviet bloc countries. Further improvements have come from the formation, under U.S. industry leadership, of the World Association of Nuclear Operators (WANO) to provide self-help among all the nuclear utilities of the world in achieving high safety standards in operation.

Major Lessons Learned

Experience sheds light on weaknesses in implementing policy-driven international cooperation. A pendulum effect can be seen over the course of events. The early enthusiasm for the transfer of atoms-for-peace technology overshadowed the central nonproliferation element of the policy objective. The formation of the IAEA and development of its institutional strength proceeded slowly, and insufficient controls were placed initially on the transfer of enrichment and reprocessing technology.

In recent times, however, prevention of weapons proliferation has overshadowed nuclear energy technology transfer. The United States had to take unilateral action at times to deter nonweapons states from nuclear weapons development. Concern over the potential for nuclear weapons proliferation through the commercial recycle of nuclear fuel led to the passage of the Nonproliferation Act of 1978, which unilaterally put restrictions on the use of U.S.-origin nuclear fuel by international nuclear utilities. Since then, the United States has substantially reduced its effort to obtain an international consensus on an adequate level of proliferation resistance in commercial nuclear fuel recycling, has banned commercial recycle in the United States, and has eliminated R&D support of commercial recycle. In parallel, the international market in nuclear power has declined sharply, causing a reduction in international commercial technology exchange. Thus the U.S. pipeline of peaceful-purpose technology is running dry, weakening the technology transfer commitment, and eroding the combined government and industry leadership in nuclear power.

The overall lesson is that a constant and consistent quid pro quo was not maintained between the benefits of the technology transfer and the obligations of proliferation control, as was the intent of the atoms-for-peace program and the NPT. The recent move of the United States toward a unilateral approach with little support of advanced reactor R&D increases this imbalance. The definition of conditions and commitments, the scheduling and funding, and the monitoring of progress should be aimed at maintaining the quid pro quo appropriate to the policy objective of future international cooperation in RD&D. This lesson will be more difficult to apply in the future because the United States no longer has the dominant technological lead it had in the 1950s and 1960s, nor is it in a fiscal position to "buy" its policy objectives. Thus the United States should reaffirm, and in some cases reestablish, its global nuclear policies and reengage with the international community in RD&D.

Status of U.S. Nuclear Infrastructure Today

Industry

With rate deregulation of electricity generation, U.S. operating nuclear plants will have to show economic superiority vis-à-vis generating alternatives in a competitive market if they are to continue to operate. Thus, the primary burden of sustaining the use of civilian nuclear power falls on the shoulders of the private sector.

Although the market is deregulated on rates, there are many other elements of regulation and related laws that have important impacts on the position of any entrant in the competition. It is unfortunate that at this point nuclear power is not competing on a level playing field because of these impacts. Nuclear power is the only generation option that has internalized in its cost structure essentially all of the safety and environmental externalities.[7] The regulations implementing the Clean Air Act have never explicitly recognized the value of the air emissions (specifically sulfur dioxide [SO_2], nitrogen oxide [NOX] particulates, and ozone) displaced by nuclear power plants. No credit is given for zero emission of greenhouse gases. Federal and state tax subsidies and preferential electricity-buying policies favor nonnuclear generation options. Safety regulation imposes costs far greater than the alternative options and out of proportion to the benefit gained because of the present unstable overprescriptive and unprioritized nuclear regulatory system. The jurisdictional overlap and conflicting regulations on radiation issues between NRC and the Environmental Protection Agency (EPA) and the reluctance to accept a safe minimal level of radiation exposure also cause instability and uncertainty that lead to higher costs.[8]

Nevertheless, recent experience indicates that the prospects are good that the nuclear utilities will successfully meet this challenge.

- U.S. operating nuclear plants are performing more effectively than ever before. Between 1980 and 1997, the average plant availability factor increased from 63 percent to 82 percent, loss of availability because of unplanned shutdowns decreased from 12 percent to 5 percent, and safety system performance as measured by the Institute of Nuclear Power Operations (INPO) increased from 70 percent to 90 percent.[9]
- Average production costs of U.S. nuclear plants dropped by 30 percent since 1988 and are now at a level of 2¢/kWh in 1996 dollars, comparable with coal-fired plants and superior to gas (3.5¢/kWh) and oil (4.2¢/kWh).[10] The total

7. Paul Portney, "Externalities of Electric Power Generation" (paper presented at Atlantic Council seminar on the future of nuclear power, Seoul, Korea, June 24, 1997).

8. *A Strategic Direction for Nuclear Energy in the 21st Century* (Washington, D.C.: Nuclear Energy Institute, December 1998).

9. *Annual Report* (Atlanta: Institute for Nuclear Operations, 1997).

generation cost of power from many nuclear plants carrying a high investment burden is higher than the above alternatives, but that burden is being alleviated by accelerated depreciation and stranded asset write-offs.

As the nuclear power infrastructure of other countries has built up, the commercial license arrangements with U.S. nuclear manufacturers have changed, limiting the benefits, the technical influence, and the market opportunities of U.S. manufacturers in those countries. These licensees have now become effective competitors in the international market. In addition, mergers of U.S. and international nuclear supplier firms have been affected, clouding the national identity of those firms. Examples of mergers are Westinghouse and BNFL, Combustion Engineering and Asea Brown Boveri (ABB), and Babcock & Wilcox (B&W) and Framatome.

Until nuclear power plants are clearly competitive on total power generation costs, there is little interest by the U.S. utilities in expanding nuclear power capacity although they would like that option to be available when that time comes. The international market for new nuclear power capacity is also in the doldrums. Thus the U.S. supplier industry has retrenched to servicing and providing fuel supplies to the present operating plants here and abroad. Overseas suppliers are competing fiercely in this limited world market.

Many other applications of nuclear technology are also being pursued, providing substantial benefits to society. A strong portion of this benefit is improved health and safety because the bulk of the applications are in medical therapy and diagnosis, food and materials sterilization, and nondestructive inspection of pressure boundaries and rotating equipment. The pace of these applications has been stopped or slowed by the lack of political and public acceptance of activities related to radiation. Of particular concern is that food sterilization, capable of saving thousands of lives annually, was held back for many years by incompetent regulatory delay and now by public relations fears.

Government

FY 1998 was a historical low point in government interest in sustaining nuclear power as an energy option in the United States. In that year the Congress eliminated DOE funding of R&D nuclear energy supply, reflecting a lack of confidence in DOE's stewardship and management of nuclear energy but also implying to the international community a lack of interest in supporting nuclear power in the United States.

With the help of recommendations from the President's Committee of Advisors on Science and Technology (PCAST)[11] and renewed support by Congress, a modest level of funding for nuclear energy supply R&D has been provided in FY 1999 in the form of the Nuclear Energy Research Initiative ($19 million) and University Nuclear Science Research Support ($11 million). PCAST also recommended that DOE provide some funding ($10 million, to be matched equally by industry) to

10. Portney, "Externalities of Electric Power Generation."

11. President's Committee of Advisors on Science and Technology (PCAST), "Federal Energy R&D for the Challenges of the Twenty-First Century" (Washington, D.C.: Executive Office of the President, November 1997).

help ensure that the present U.S. nuclear power capacity is maintained in view of its capability of producing electricity without air pollution and greenhouse gas emissions. DOE proposed such support in its FY 1999 budget (the Nuclear Energy Plant Optimization Program), but Congress rejected the proposal.

DOE's FY 2000 nuclear energy supply R&D budget proposes $25 million for the Nuclear Energy Research Initiative, $11.3 million for university nuclear science research support and $5 million for nuclear energy plant optimization.[12] In comparison with the DOE FY 2000 budget proposal of $41.3 million for nuclear energy supply R&D, DOE has budgeted $1,240 million for energy efficiency and renewable energy and $317 million for fossil fuels R&D.[13] Compared with FY 1999, funding for nuclear energy supply is increased by $11.3 million, energy efficiency and renewable energy is increased by $270 million, and fossil energy is decreased by $239 million.

This allocation is questionable in light of the fact that fossil and nuclear fuels are the two largest sources of U.S. electricity generation today, producing 50 percent and 20 percent respectively of the nation's supply, while the contributions of non-hydroelectric renewable sources (solar, wind, geothermal, and biomass) are negligible. Furthermore, the Energy Information Administration estimates that non-hydroelectric renewable energy will contribute only 5.5 percent of the nation's electricity by the year 2015, assuming passage of renewable-portfolio-standards legislation that would subsidize renewables at an annual cost to the ratepayers of between $1.4 and 3.7 billion between 2005 and 2010.[14]

Nuclear technologies are also funded for a variety of purposes other than nuclear energy supply. The Congress, recognizing the benefits of radioisotope applications, is funding DOE at approximately $20 million annually for the production of isotopes. A related program has also been funded at a level of $20 million annually to obtain a better scientific understanding of the health effects of low-level radiation.

Billions of dollars are being provided annually to DOE for spent fuel disposition, for cleanup of defense facilities, for U.S. participation with Russia in improving stored nuclear weapons security and disposing of excess U.S. and Russian nuclear weapons materials, for the nuclear weapons stockpile stewardship program, and for maintaining or decommissioning test facilities. Yet the pace of these programs is slow. Although the national security programs continue to be the mainstay mission of DOE, the nuclear infrastructure—people, expertise, and facilities—supporting them is being downsized even as new demands driven by stockpile stewardship and weapons proliferation prevention are added to the

---

12. William Magwood, "The FY 2000 Nuclear Energy, Science, and Technology Budget Request: Lighting the Way for the New Millennium" (paper presented to the DOE Nuclear Energy Research Advisory Committee [NERAC], Washington, D.C., Energy Information Administration, March 29, 1999).

13. U.S. Department of Energy, *FY 2000 Congressional Budget Request—Science, Security, and Energy: Powering the 21st Century,* DOE/C R-0059 (Washington, D.C.: GPO, January 1999).

14. S. Sitzer, "The Outlook for U.S. Nuclear Power through 2020" (paper presented to the DOE Nuclear Energy Research Advisory Committee [NERAC], Washington, D.C., Energy Information Administration, March 29, 1999).

national agenda. The nuclear navy has substantially downsized and is slowly moving toward a small fleet and infrastructure, evolving into a fleet-stewardship mode that is analogous to the weapons stockpile stewardship program.

The technologies being developed by the weapons and naval nuclear propulsion laboratories are of significance to global management of nuclear materials in future commercial nuclear power activities. Examples are

- measurement techniques and telecommunication tools applicable to safeguards, materials monitoring, and protection,
- nuclear data on potential proliferant nuclei,
- detector calibrations for archival diagnostic data analysis and the integral evaluations of nuclear material properties, and
- technologies capable of verifying weapons-usable HEU and plutonium inventories applicable to the transparency requirements of commercial fuel inventories.

The nuclear energy organizational structure in DOE is fragmented into many independently operated offices: Office of Nuclear Energy, Science, and Technology; Office of Nonproliferation and National Security; Office of Civilian Radioactive Waste Management; Office of Environmental Management; and Office of Fissile Materials Disposition. This is in sharp contrast to the centralized approach taken in the past with the Atomic Energy Commission. Each office must look to the under-secretary-of-energy level for coordination, adding greatly to the coordination responsibilities that cover all of the DOE scope at that level. This organizational arrangement and the slow pace of the nuclear energy and legacy programs portray a picture of disarray that gives credence to the call of Senator Pete Domenici for an integrated U.S. approach to nuclear energy.[15]

Universities

The weakening trend seen in industry and the government is mirrored in the U.S. universities, another key measure of national strength. Many universities have merged their nuclear engineering departments into other departments or have completely eliminated them. At this time there are only 32 schools in the country with degree-granting programs and only 20 with nuclear engineering departments.

The population of matriculating students has declined drastically. Since 1992, the total number of nuclear engineering undergraduate students in the United States has decreased by 62 percent, master's candidates by 44 percent, and doctorate candidates by 30 percent.[16] The ratio of foreign students to domestic students pursuing graduate degrees in nuclear science and engineering has increased from 2:10 to more than 7:10 between the 1970s and the 1990s.[17] More of these foreign stu-

---

15. Pete V. Domenici, "A New Nuclear Paradigm," (paper presented at the inaugural symposium of the Belfer Center for Science and International Affairs, Harvard University, Cambridge, Mass., October 1997).

16. J.P. Friedberg and G.S. Was, "Nuclear Engineering in Transition: A Vision for the 21st Century" (paper presented to the DOE Nuclear Energy Research Advisory Committee [NERAC], Washington, D.C., Energy Information Administration, March 30, 1999).

dents are returning to their own countries to participate in their national programs compared with the past when most obtained superior jobs in the United States. R&D funding, from both government and private sources, has declined significantly.

Overall Assessment

On all fronts—industry, government, and university—the technical strength of the United States in nuclear energy supply and nuclear technology is continuing to weaken, making it more difficult to provide knowledgeable and credible leadership to support the global and largely bipartisan nuclear policies the U.S. espouses. The United States has lost the lead in many areas of nuclear energy technology, notably, test facility capability, nuclear plant fabrication and construction, and the nuclear fuel cycle. The 1995 *National Critical Technologies Report* issued by the administration's Office of Science and Technology Policy concluded that "... the United States is likely to fall behind in next-generation reactors because of large funding cuts for reactor R&D."[18] Events since 1995 make for an even more gloomy assessment. A recent signal of this deterioration in capability is the decision of DOE to award the first-phase contract for burning excess weapons plutonium in U.S. reactors to a consortium led by the French firm Cogema and including a Belgian firm and the French nuclear supplier, Framatome. The only U.S. participants are a subsidiary of Framatome, the utilities that own the U.S. reactors, and Stone & Webster, the U.S. architect engineer.

The combined leadership of the U.S. government and industry has deteriorated, weakening the negotiating ability of the United States to build a fully effective international nuclear weapons control regime and to continue to ensure safe and proliferation-resistant nuclear power throughout the world. In the same time frame, existing and emerging nuclear power nations have slowly weaned themselves of dependence on U.S. support, goods, or services. Further, the industry has been left alone to compete in a rate-deregulated, but adversely tilted, competitive market.

Present U.S. Global Nuclear Policies

Policies Governing Nonproliferation of Nuclear Weapons and the Safety and Environmental Impact of Nuclear Power

Although the present U.S. policy with respect to the domestic role of nuclear power is ambiguous and subject to political controversy, certain U.S. international nuclear policies have been constructively formulated and enjoy substantial bipartisan support. They largely originate from the NPT and have been bolstered by many other

17. U.S. Senators Larry E. Craig, Pete V. Domenici, Richard J. Durbin, Dirk Kempthorne, Jon Kyl, and Frank Murkowski (letter written to Secretary of Energy Federico F. Peña, July 30, 1997).

18. J.H. Gibbons, *National Critical Technologies Report* (Washington, D.C.: Office of Science and Technology Policy, March 1995).

agreements, conventions, and implementing activities of the IAEA.[19] The present administration

- calls for international control and tight security of weapons-usable fissile materials to discourage an increase in the number of weapons states and to prevent access by terrorists and rogue states (NPT and IAEA Convention on Physical Protection of Nuclear Material),
- fosters exchange of peaceful use technology with the nonweapons states (NPT),
- espouses a reduction in nuclear arms by the weapons states (NPT and START II) and the cessation of nuclear weapons testing (Comprehensive Test Ban Treaty),
- supports the establishment of high international standards of nuclear power plant safety and radioactive waste management and strong national safety regulation (IAEA Convention on Nuclear Safety and IAEA Convention on Safety of Spent Fuel and Radioactive Waste Management), and
- assists the former Soviet bloc countries in improving the safety of their nuclear operations and waste management (Nunn–Lugar and related legislation).

Problems of Implementation

The problem is not with these policies but with their implementation. For example, the improvement of the security of excess weapons materials and the disposition of excess weapons plutonium is moving slowly in light of the risks involved. The effort to improve IAEA nonproliferation surveillance needs stronger support from the member nations. International standards need greater refinement, particularly in the physical protection and waste management areas, as well as a higher level of international compliance. The international assistance programs need strengthening, both in funding and in focus. Although these problems should not be laid at the doorstep of the United States alone, there are many ways in which the United States can improve the implementation processes. The National Academy of Sciences, CSIS, and the Atlantic Council have all completed studies that recommend ways for such improvement.[20]

Policy on Recycle of Nuclear Fuel

U.S. policy governing the commercial recycle of nuclear fuel has gyrated through successive administrations and has been a source of controversy among other coun-

19. Fischer, *History of the International Atomic Energy Agency.*

20. Reactor Options Panel of Committee on International Security and Arms Control, National Academy of Sciences, *Management and Disposition of Excess Weapons Plutonium: Reactor Related Options* (Washington, D.C.: National Academy Press, 1995). See also "An Appropriate Role for Nuclear Power in Meeting Global Energy Needs," (Washington, D.C.: Atlantic Council of the United States, February 1999); *Disposing of Weapons-Grade Plutonium: A Consensus Report of the CSIS Senior Policy Panel on the Safe, Timely, and Effective Disposition of Surplus U.S. and Russian Weapons-Grade Plutonium* (Washington, D.C.: CSIS, March 1998); "Proceedings of CSIS Conference on Global Nuclear Materials Management," (Washington, D.C.: CSIS Energy and National Security Program, December 4, 1998).

tries with nuclear power capabilities. The present U.S. administration policy prohibits commercial recycle in the United States and discourages it overseas but does not interfere with the commercial recycle activities of its allies. For all practical purposes, the United States has disengaged itself from the international community on this issue except for expressions of concern, with which IAEA concurs, for the buildup of commercial separated plutonium stocks.

Present U.S. Global Policy Direction

Government Leadership

A recurrent theme in the studies cited above is the need for national leadership on policy implementation at the top of the administration.[21] Simply stated, a prerequisite for global leadership is national leadership. There are competent, dedicated DOE (as well as Department of State and other departments) managers of the policy implementation programs who are frustrated at every turn by the management people above them who keep shifting their interpretation of the ambiguities that are caused by lack of clear leadership. This difficulty is further aggravated by antinuclear sentiment among some higher-level administration figures. Although other actions, discussed below, are important in restoring U.S. global leadership, their impact will be blunted without leadership at the top of the administration.

Leadership on nuclear energy issues at the cabinet level is also essential, and Secretary of Energy Bill Richardson has moved in that direction in his early days in office.[22] To assist in achieving that leadership, the secretary should be given clear lines of authority to determine courses of action that are not impeded by others at the cabinet level and higher. One option would be a cabinet-level nuclear policy board, representing Commerce, Defense, Energy, and State. DOE organizational changes could also be made to centralize better the nuclear programs, perhaps by establishing an executive director for nuclear energy to whom all the nuclear offices report.

Strengthening of DOE nuclear energy staff is important for the future because many experienced staff have retired or are near retirement. A vehicle proved in the past is an agency internship program in which young, high-potential people are hired on the basis that seniority will not be rigorously applied to them in the event of a reduction in force.

Consideration should also be given to some degree of centralization at the congressional level. Increased understanding and prioritization of the overall energy R&D portfolio would be achieved by consolidating all energy research funding in both the House and the Senate under the Subcommittees (of the House and Senate Appropriations Committees) on Energy and Water Development. Jurisdiction over energy research is currently shared between the Energy and Water Development

21. Ibid.

22. Bill Richardson, U.S secretary of energy (message from President Bill Clinton delivered to the general conference of the IAEA, Vienna, Austria, September 21, 1998).

Subcommittees and the Interior Subcommittees of the Appropriations Committees in both houses.

Implementation of Bipartisan Global Policies

NUCLEAR ARMS REDUCTION AND FISSILE MATERIALS CONTROL. The effectiveness and pace of implementation of the bipartisan global policies on fissile materials control and nuclear arms reduction, which are essential to enhancing public perceptions of nuclear power, should be accelerated.

The present efforts to burn excess HEU in reactors and the planning for disposal of excess weapons plutonium by reactor burning and immobilization are commendable.[23] All of the elements of the plan for excess weapons plutonium disposition are now in place but the rate of progress is not commensurate with the urgent need and can be greatly improved by following recent CSIS recommendations.[24]

Capabilities for controlling international fissile materials controls should be improved as follows:

- Strengthen support for the IAEA safeguards program, including the 1997 additional protocol, that provides for unannounced inspections;
- Strengthen support for the Convention on Physical Protection of Nuclear Material and its accompanying guidelines,[25] recently expanded to protect against sabotage; and
- Develop an international consensus on adequate proliferation resistance in commercial power, from both a technical and an institutional standpoint.

INTERNATIONAL NUCLEAR POWER PLANT SAFETY. The major efforts by Organization for Economic Cooperation and Development (OECD) nations and the IAEA since the Chernobyl accident to assist the former Soviet bloc countries in improving the safety of their nuclear power facilities have resulted in substantial improvements in operational safety and a modest level of equipment upgrading. Yet the funding of the overall effort has hardly been commensurate with the need. Many of the Soviet-designed plants are not up to the safety standards of the OECD countries but are continuing to operate because the electric power requirements of the communities they serve cannot be met any other way in the near term. The only exception to this pattern has been the decision to shut down and decommission the nine Soviet-designed Griefswald and Rheinsberg nuclear power plants in eastern Germany, made possible by the German reunification and the ability of western Germany to supply the electricity. At the present pace, the hazards of inadequately safe nuclear power plants in the former Soviet bloc will cloud the worldwide nuclear power industry for many years to come.

23. U.S. Department of Energy, "Record of Decision for the Storage and Disposition of Weapons-Usable Fissile Materials Final Programmatic Environmental Impact Statement," (issued in accordance with the National Environmental Policy Act) (Washington, D.C., GPO, January 14, 1997).

24. *Disposing of Weapons-Grade Plutonium.*

25. "The Physical Protection of Nuclear Materials and Nuclear Facilities."

Besides limited funding, there have been insufficient coordination and inadequate planning of these cooperative safety efforts among the West European and Asian countries and the United States. The problem is caused by several factors. National budget austerity has restricted the level of funding. Too much of the limited funds have been spent on analysis and evaluation instead of providing specific equipment, systems, and operational and maintenance procedures. The potential availability of such funds to national participants has also created competition for the funds, not always to the benefit of the priority safety objectives. The desire to maximize the assignment of contract dollars to the individual funding nations interferes with a coordinated planning process and the setting of safety priorities. An overall concern is the lack of a consistent drive, on the parts of both the former Soviet bloc countries and the OECD countries, to solve the problem with a strong policy position and adequate funding. The erratic shifts by Ukraine and the assisting nations on whether and when the Chernobyl units will be shut down and how their capacity will be replaced is a case in point. The situation calls for a policy-driven, international, project-oriented collaboration with focus on the tangible improvement of the safety of the Soviet-designed reactors and their associated facilities and waste storage.

Substantial effort should be continued to improve international safety standards for reactors and their associated facilities. The 1996 IAEA Convention on Nuclear Safety is a means of stimulating each nuclear power nation to establish a sound regulatory framework for national reactor safety that is consistent with IAEA safety principles. Implementation of the peer review process delineated in the convention deserves sustained support by the IAEA member states.

INTERNATIONAL SPENT FUEL AND RADIOACTIVE WASTE MANAGEMENT.

There are compelling reasons to engage in collaborative international research, development, and deployment efforts to address the final disposition of spent fuel and high-level radioactive waste. Resolution of this issue is a prerequisite to continued expansion of nuclear power capacity. The technology encompasses a broad range of disciplines and the deployment costs are high; thus sharing of costs and technical talent is important. Because the licensing processes are still in their infancy, there is great merit in having a consistent international regulatory framework early on, including an international consensus on retrievable storage and repository security and nonproliferation standards. The 1997 IAEA joint convention on the safety of spent fuel management and the safety of radioactive waste management is an important vehicle through which to gain such consensus. The convention was signed by the United States and should be ratified promptly.

International collaboration beyond the level of study and R&D exchange is needed. In particular, several recent proposals urge the development of regional interim storage facilities and repositories.[26] A key feature of two of these proposals is to involve Russia in providing interim storage of spent nuclear fuel for itself and other countries, using the revenue earned to improve the security of their excess nuclear weapons materials. Specific recommendations on developing regional storage facilities are being provided by Task Force II, and an assessment of the concept is provided in a DOE-sponsored study.[27]

Redirection of Domestic Nuclear Energy Policy

The need for redirection. The long-term global energy demand, the concerns about air pollution and global warming, and the potential of high-conversion nuclear fuel systems to open up a vast untapped source of nuclear fuel over the next century call for a more effective U.S. domestic nuclear energy policy.

Nuclear energy is unique in providing an essentially carbon-free energy source. Nuclear power plants have reduced air pollution in the U.S. electric power industry, avoiding the emission of about 80 million tons of SO_2 and 34 million tons of NOX. Nuclear power plants are responsible for 89 percent of all the carbon dioxide emissions avoided by U.S. electric utilities between 1973 and 1995, eliminating the emission of 1.9 billion metric tons. In 1996 alone, emissions would have been 147 million metric tons higher in the absence of nuclear energy.[28] No strategy to stabilize greenhouse gas emissions can be credible without a significant nuclear energy component.

Recent projections confirm that, if the potential energy in uranium can be realized by transmutation to plutonium, the nuclear fuel resource base is three times larger than the total fossil resource base and extends the effective life of uranium supplies 50-fold.[29] A further several-fold incremental increase in the nuclear fuel resource base could be achieved by transmuting thorium to uranium 233. Nor has the need subsided for a nuclear power contribution to the massive requirements for future world electricity production. Under high- and low-growth scenarios, electricity is expected to become the principal end-use energy, growing from 30 percent to 60 percent of primary energy. New electric generation capacity requirements could reach 10,000 GW by 2050, requiring 1000 MW of new capacity on the average of every 4 days to meet the increased demand and replace existing stock.

What this picture calls for is an unambiguous U.S. policy supporting the preservation of the nation's existing nuclear power capability, the enhancement of that investment by extending the licenses of the present plants, and the option to expand that capability to meet future power needs.

Actions to support redirection. Support of such a policy calls for the following actions by both government and industry:

---

26. See Chauncey Starr, "Proposal for an International Monitored Retrievable Storage Facility (IMRSS)" (paper presented at Workshop for Securing the Nuclear Future, Los Alamos National Laboratory, 1995) and Atsuyuki Suzuki, "Proposals for Regional Storage of Spent Fuel" (paper presented at the CSIS Conference on Global Materials Management, Washington, D.C., December 1998 and at the Okinawa Energy Business Forum, Okinawa, October 1998). See also Thomas B. Cochran and Christopher E. Paine, "Proposal for Augmenting Funding for the Disposition of Russian Excess Plutonium" (briefing for U.S. Arms Control and Disarmament Agency [ACDA] Director's Advisory Committee, at the Natural Resources Defense Council, Washington, D.C., November 18, 1998).

27. Lewis A. Dunn and Stephen Carey, "Internationalizing Spent Fuel Storage: Concepts, Issues, and Options" (working paper prepared for Office of Arms Control and Nonproliferation, Office of Nuclear Safety and Security, DOE; and Office of the Secretary of Defense) (San Diego, Calif.: Science Applications and Technology, Inc., February 25, 1998).

28. *A Strategic Direction for Nuclear Energy in the 21st Century.*

29. N. Nakicenovic, A. Gruebler, and A. McDonald, eds., *Global Energy Perspectives to 2050 and Beyond* (Cambridge: Cambridge University Press: 1998).

- Safety and environmental regulation, essential for all generation alternatives, should be reevaluated to establish uniformity in public safety and health protection. Risk assessment methodology has now reached a capability that can establish a common level of acceptable risk for all alternatives, including the entire generating cycle: mining, fabrication, construction, fueling, operation, waste management, and decommissioning.
- Overlapping regulatory jurisdictions should be eliminated by refinement of the responsibilities of the agencies involved.
- Present efforts by the NRC to improve the effectiveness of regulation should be pursued vigorously:
  - License renewals should be processed expeditiously; and
  - The move to a performance-based, risk-informed process for regulation, inspection, and enforcement should be accelerated and brought to a stable conclusion.
- Substantial improvements in national spent fuel management are needed:
  - Legislative changes in the Nuclear Waste Policy Act to provide central interim storage of spent nuclear fuel, a safe transportation system, and ultimate disposal;
  - Enforcement of the federal obligation to accept used nuclear fuel; and
  - Accelerated progress on the Yucca Mountain repository, based on the recently completed site viability assessment.
- Actions to increase the viability of low-level radioactive storage are needed:
  - Maintain access to existing capacity and promote access to new capacity;
  - Streamline approval for alternative disposal requests; and
  - Exempt commercial mixed waste from dual regulation under the Resource Conservation and Recovery Act.
- Define low-level radiation health effects realistically:
  - The low-dose radiation research program that was initiated by the Congress in FY 1999 should be pursued vigorously by the Office of Science and the Office of Environmental Management;
  - Recycling standards for radiation-contaminated metals should be defined;
  - Allowable residual levels of radiation at decommissioned sites should be defined; and
  - Emission credits should be allocated based only on unit power output.

Long-term support of redirection. These actions are urgently needed to support existing U.S. nuclear power capacity and nuclear technological capability. They are also essential as a foundation for the development and deployment of future advanced systems. The need to deploy advanced systems, which can dramatically expand nuclear fuel supply by transmutation, has been delayed from the early to the middle decades of the coming century, however. Thus there is time to address the safety, environmental, and proliferation control issues of the commercial, high-conversion nuclear fuel cycle well before deployment.

The present perception held by many countries is that the present U.S. policy, although allowing plutonium recycle activities by its allies, fosters an effort to kill nuclear fuel recycle throughout the world. As a result, there is little dialogue or cooperation on addressing proliferation resistance in commercial recycle. The U.S. policy should be amended to state that long-term energy needs, air pollution, and global warming considerations require the prudent development of high-conversion nuclear fuel systems and that the United States is willing to cooperate with other nuclear power countries in their development. Such an amendment would further state that U.S. cooperation is contingent upon a serious program to establish international standards and institutional mechanisms to ensure that diversion to nuclear weapons use will not result from commercial deployment.

Without such involvement by the United States, it is hard to see how the United States can realize its nonproliferation policy objectives in future nuclear power plant deployment. An amended policy approach of this nature could stimulate the industry to cooperate and participate in developing more effective nonproliferation standards. As suggested in the Phase I CSIS Global Nuclear Materials Management Forum, there is a leadership role for industry here but it will not happen without U.S. government encouragement.[30]

The primary implementing action of such an amended policy would be the initiation of strong international R&D collaboration. A key early task of such collaboration is to develop an international consensus on an adequate level of proliferation resistance for the deployment of commercial, high-fuel conversion reactors and fuel recycling systems, both from the technical and institutional standpoint. Since the need for broad deployment is delayed, there is time to engage in R&D on promising recycle or once-through advanced high-conversion systems. There is merit in a clean-slate approach to developing this consensus, as recommended in the Nuclear Energy Research Initiative Workshop.[31]

A Partnership Issue

A key focus of the above recommendations is to improve international cooperation in the pursuit of U.S. global nuclear policies with strengthened U.S. leadership.

---

30. John J. Taylor, "A Leadership Role for U.S. Industry in the Emerging Nuclear Era" (paper presented at the CSIS Conference on Global Nuclear Materials Management, Washington, D.C., December 4, 1998).

31. U.S. Department of Energy, "Summary Report of the Nuclear Energy Research Initiative Workshop: An Assessment of Research Opportunities for Nuclear Energy, Science, Technology, and Education" (Washington, D.C.: GPO, June 1998).

Without a significant improvement in cooperation domestically, however, U.S. international leadership will be inhibited. There are significant antagonisms between government and industry over regulatory issues. There is still a major controversy among government, industry, and the environmental community over the future direction of nuclear power. There are also rivalries among the technical interests in various R&D organizations as to the priorities and allocations of funds for nuclear energy supply R&D. Contention over such issues as radioactive waste management greatly raises the cost of remedial actions, reducing the money available to address these key global nuclear policies. Unless these antagonisms are ameliorated, there is not much chance of gaining public understanding and acceptance of any of these nuclear policies, old or new. Without such acceptance, the chances are low of sustaining the long and costly effort needed to achieve the global policy goals.

The Decision Makers' Forum was an important first step forward in developing consensus among all of these parties on national planning for a new paradigm for nuclear energy.[32] Continued follow-up is needed to assess the level of progress, adjust direction as needed, and broaden the participation of the environmental community. The formation of the DOE Nuclear Energy Research Advisory Committee (NERAC) provides a forum representing these many interests through which consensus on R&D priorities can be developed. Strong representation of both the environmental and the academic communities on NERAC is a fresh and promising move.

In this second nuclear era, the disparate interests of the contending parties are tending to converge around similar objectives as the complexity of the post–Cold War situation emerges and the global nature of environmental protection becomes more and more apparent. A good example of this is the initiative by the environmental community to foster interim spent fuel storage in Russia in the interests of proliferation control.[33] Opportunities should be sought for similar creative initiatives that serve the common objectives of the government, environmental community, and industry.

Conclusions

The combined leadership, talents, and resources of the U.S. government and industry forged a fundamentally sound technical and institutional foundation for the control of nuclear weapons materials and the utilization of nuclear energy for peaceful purposes in the early decades of the nuclear age. Since that foundation is by no means perfect and the nuclear world is changing rapidly, much needs to be done to build on, and further improve, that foundation for the new nuclear era. This is a compelling reason to reinvigorate the combined U.S. leadership that will elicit the cooperation of other countries, just as was done in the first, and less complex, nuclear era.

---

32. "Report of Decision Makers' Forum on a New Paradigm for Nuclear Energy."
33. Cochran and Paine, "Proposal for Augmenting Funding."

The global nuclear policy established by the United States in the first nuclear era to prevent weapons proliferation and open up the benefits of nuclear technology has not changed, but the United States has moved to a more unilateral, rather than collaborative, approach. Other nations have attained excellence in nuclear technology while the technological strength in U.S. universities, government, and industry has waned and its national nuclear policy to harness nuclear power for the benefit of the nation has weakened.

To correct these trends, the improvements recommended above in implementing existing U.S. global policies on fissile materials control, nonproliferation, and nuclear facility safety should be followed. Of equal importance, a major change in national nuclear policy should be made; it should follow the broad recommendation of the Decision Makers' Forum sponsored by the Senate Nuclear Issues Caucus on a New Paradigm for Nuclear Energy:

> ...create a broad-based nuclear mission that advocates a viable commercial sector, produces top-quality scientists and engineers, develops options and techniques for disposal and recycle, and funds reactor and fuel cycle R&D as an integrated component of an overall energy portfolio.[34]

The United States is at a crossroad in nuclear energy. Will it reengage with the international community, exerting effective leadership through redirected domestic policy and improved implementation of its bipartisan global policies? Or will it drift out of leadership that it can no longer afford to buy, allowing the nuclear policies of other nations to dominate?

---

34. "Report of Decision Makers' Forum on a New Paradigm for Nuclear Energy."

Part Two

Introduction

This project, Global Nuclear Materials Management, had its beginnings on December 4, 1998, when the Center for Strategic and International Studies (CSIS) convened an all-day conference to define those challenges emerging from the new nuclear era. Four panels were assembled to address topics chosen to reflect both domestic and international concerns:

- The end of the Cold War and the emerging nuclear era,
- Nuclear infrastructure—national and international trends,
- The Nuclear Non-Proliferation Treaty—30 years later, and
- U.S. nuclear policy in the post–Cold War era—what's next?

These presentations identified both challenges and opportunities for the United States, in partnership with the world community. In brief, the panelists together isolated five issues they believed warranted full examination and the development of policy recommendations. These issues were, in no particular order of priority:

- Funding nuclear security,
- The concept of an international spent fuel storage and disposal facility,
- Commercializing the excess nuclear infrastructure of Russia,
- Nuclear materials transparency, and
- The U.S. infrastructure in the emerging nuclear era.

CSIS then formed five task forces—one for each of the issues identified at the December 4 conference—and established a coordinating committee to pull together the efforts of these task forces in an executive summary[1] and, where appropriate, look beyond the original task force individual assignments. Finally, a senior policy panel was appointed to review and approve the work of the coordinating committee.

The findings and recommendations of these five task forces, together with the executive summary, were presented at a CSIS conference on July 22, 1999. To balance the agenda, Senators Pete Domenici and Bob Kerrey offered a view from Capitol Hill on the issues under discussion, while DOE Under Secretary Ernest Moniz presented the administration's position. Senator Domenici noted that although Congress is taking an increasing interest in nonproliferation that has generated new legislative proposals, these proposals contained too many loopholes and compromises. Senator Kerrey, in his prepared remarks, stressed that it is not the

1. The executive summary is on page xiii.

fear of a planned nuclear attack that threatens us now. Instead it is nuclear proliferation, political instability, and decaying infrastructures that offer the greatest threats. Secretary Moniz, while acknowledging that progress on the many nuclear-related issues appeared slow, nonetheless said there had been successes achieved in terms of cooperation between the United States and Russia on military issues; in terms of the HEU purchase agreements; and under the MPC&A program, which has secured significant amounts of weapons-grade material.

William Potter, as the concluding speaker, offered his view of the future, with that view centering on the fragile state of the Nuclear Non-Proliferation Treaty, the growing potential for a major accident at a civilian nuclear power reactor in Russia, and the need for training the next generation of U.S. leadership in nuclear matters.

James Schlesinger, in his keynote address, stressed that the disciplines of the Cold War are now gone, the amount of nuclear-related information is much greater, and the temptation to reach for nuclear weapons is greater. The United States faces a formidable challenge, and we should not hope for total success.

Part Two of this report presents the remarks of these speakers in full. What they had to say adds considerable strength to the senior policy panel's findings and recommendations while it brings other viewpoints into play.

CHAPTER 6

A View from the Hill

Remarks by Senator Pete V. Domenici and Senator J. Robert Kerrey

Senator Domenici:

I'm glad to be with you today for your discussions on global nuclear materials management. Your conference chairman, my colleague Sam Nunn, is superbly qualified to lead these discussions.

Control of nuclear materials is a very real issue for the world. For that reason, when I initiated a national dialogue on nuclear issues at Harvard in October 1997, this subject was a key part of my overall concerns. And, like most nuclear issues, your subject today has an interplay of both military and civilian perspectives.

From a military perspective, nuclear materials raise serious proliferation issues. Nations with nuclear weapons must carefully protect whatever minimum stockpiles of these materials they decide to maintain. And as future arms control regimes come into play, we can all hope that the quantities of materials needed by each country for weapon programs can continue to drop. That will lead to still larger surpluses of nuclear materials. It is in the vital interest of all nations to ensure that these surplus materials are not diverted into military applications by other nations. This requirement stresses safeguard regimes to their utmost.

The threat of proliferation is especially prominent after the recent nuclear tests in India and Pakistan. While we can hope that they are the last entrants into the nuclear weapon club, careful attention to management of nuclear materials may add some insurance to these wishes. There remain signs that nations like North Korea, Iran, or Iraq may be seeking to join this club.

Programs like Materials Protection, Control, And Accounting offer one of our best opportunities to block new nuclear states by avoiding diversion of weapons materials. I've supported this program throughout its history. The Senate just recently increased the administration's proposal for next fiscal year by $20 million, to a total of $165 million. With these increased funds, progress on new opportunities with Russian naval nuclear fuel should be possible.

Military materials also can have significant environmental consequences that also influence neighboring countries. Current international efforts to assist Russia with its naval fuels are a good example of the type of global cooperation that is essential.

From a civilian perspective, many nations have growing stocks of spent nuclear fuel. No nation has a complete strategy in place to deal with this issue. While nuclear energy offers immense environmental benefits with its emission-free performance, environmental concerns will remain as long as spent fuel strategies remain tentative. Some nations, especially France, are far ahead of the United States in progressing toward a complete strategy. We have a great deal to learn from their experiences.

Furthermore, it is well known that civilian fuel can present a military threat if a nation chooses to divert materials from the civilian cycle. That concern has to be integrated into overall global materials management strategies.

There are some hopeful signs for progress on civilian spent fuel and a growing recognition in Congress that actions must be taken to inject more certainty into the time scales for removal of spent fuel from utility sites. We've watched the acceptance date for Yucca Mountain slip from the planned 1998 date to an earliest possible date of 2010. And, while last year's viability assessment on the repository was positive, I doubt that anyone is lulled into believing that this action makes a 2010 date very secure.

An interim waste storage bill was introduced into the Senate and discussed extensively in the Energy and Natural Resources Committee. That bill would have allowed waste acceptance in 2003 but, as usual, the administration opposed the bill.

Senator Frank Murkowski has now presented another bill, which strives for a compromise on issues separating the administration and Congress. For example, the new bill couples the timing for an early receipt facility at Yucca Mountain to the issue of the construction license for the repository. Thus, a slip in the Yucca Mountain schedule also delays the early receipt facility. The new bill also allows the federal government to take title to waste at utility sites, which has the effect of leaving the waste in place around the country longer than I'm comfortable with.

This new bill has now garnered significant bipartisan support and was reported out of committee by a 14-6 margin. I supported this new bill because we must move forward on nuclear waste issues, even though the compromises go further than I'd like.

Even with major compromises, there is still some Democratic opposition to the new bill, mainly over the roles of the EPA (Environmental Protection Agency) and NRC (Nuclear Regulatory Commission) in setting radiation protection standards. Some argue that the EPA's decision should be further reviewed by a cabinet-level panel. I've argued in favor of the NRC assuming this role. The NRC has the expertise to set these standards and is not driven by political agendas. In fact, having a cabinet review of an EPA decision just creates a double dose of political impact.

The new bill includes a feature that I proposed, a new Office of Spent Nuclear Fuel Research to develop options for an improved future national spent fuel strategy. This research should provide tools for a future Congress to decide whether spent fuel should be treated only as waste destined for a repository or for some form of energy recovery through recycle. It should provide solid data for evaluating options for the type of waste entering a repository and determine if risks and costs justify treatment of the waste to reduce its toxicity.

The office will explore advanced reprocessing technologies as well as transmutation by both accelerators and reactors. The research will use international cooperation, and I know from personal contacts that many countries are eager to cooperate on development of international approaches to spent nuclear fuel.

Accelerator transmutation of waste, or ATW, is under limited study now, with a road map being developed by the DOE. The current Senate bill funds ATW at $15 million in the next fiscal year.

Transmutation has major economic uncertainties, but it may offer an important alternative for our future spent fuel strategy. With transmutation, the residual energy in spent fuel can be recovered and the toxicity of the resulting waste vastly reduced. Transmutation could dramatically impact the difficulties associated with siting a repository or alter the design of future repositories.

Returning to the military side of your issue, I've closely followed and encouraged programs to move highly enriched uranium, or HEU, and weapons grade plutonium into forms that would not permit future weapons use. At least for the moment, the HEU agreement in on track, thanks to major intervention by Congress last year. As you know, the HEU agreement covers enough material for about 20,000 weapons.

I understand that the State and Energy Departments are making good progress on negotiations to dispose of 50 tons of weapons-grade plutonium from both the United States and Russia. This 50 tons is enough for at least 6,000 bombs. The negotiations are on target for a September agreement, and I might note that the September date is extremely important. We need agreement then in order to influence support in appropriations for next year. I'm also pleased to learn that international negotiations are making some progress in encouraging other nations to accept some Russian MOX fuel, which would allow increased disposition rates.

Let me close with a note of concern about the deteriorating nuclear expertise in United States. This is evidenced by plummeting enrollments in nuclear engineering, as well as by decreasing numbers of universities that even have a nuclear engineering department. As just one statistic, the total number of undergraduate students enrolled in nuclear engineering deceased by 62 percent since 1992. And these decreases will continue until students see a real future for nuclear technologies.

Whether we talk about nuclear energy or management of nuclear materials, we need highly trained specialists in nuclear technologies. That's why Congress started the Nuclear Energy Research Initiative last year, and the Senate has proposed to expand it to $25 million next year. The Senate also proposed to start a Nuclear Energy Plant Optimization Program next year for $5 million that should help to rebuild some of the nation's nuclear capabilities.

A number of congressional initiatives may be on the horizon that have an impact on nuclear issues. For example, there may be further reform in the Nuclear Regulatory Commission in addition to the significant progress that the NRC has recently made in streamlining their own procedures.

I've already mentioned pending bills for spent fuel strategy. That is an area that cries out for the administration to recognize the strong concerns of Congress and taxpayers and allow real progress.

Electricity deregulation may be considered, and the form of that bill could affect nuclear issues. For example, the administration proposes a 7.5 percent set-aside for renewable energy sources by 2010. I think a far more sensible approach would be to create a level playing field by requiring a low-emission portfolio, at a far higher percent, perhaps 30. Such a change would benefit the taxpayers in many ways, emissions would be reduced, and the market could choose the right mix of energy sources.

In addition, any deregulation bill should require that EPA issue future emission credits based only on the output of a facility and stop determining credits based on the amount of fuel that is burned. The current practice doesn't create effective incentives for low-emission sources.

Progress on global management of nuclear materials is one of the most critical elements that can allow us to derive full benefits from nuclear technologies. Nuclear energy is already of immense importance to the United States and other nations from energy and economic security perspectives. We need to keep nuclear energy as a viable option for future energy needs, but we'll never realize that option if we fail to meet the challenge of managing nuclear materials.

Senator Kerrey:

I would like to thank CSIS for the opportunity to address today's conference on global nuclear materials management. I admit I was a bit concerned when I first received the invitation from Richard Fairbanks and saw this part of the program had the rather lofty title, A View from the Hill. To me, this implied a certain level of insight and clarity of thought not commonly expected from Congress. However, I will do my best to share with you my thoughts on the issue of U.S. leadership in the emerging nuclear era.

As I opened up the paper this morning and saw the North Korean regime is yet again preparing to test launch a long-range ballistic missile, the timeliness of today's conference became even more apparent. The stability of the entire East Asia region could be in jeopardy as a result of a North Korean missile test. North Korea is one of the most backward countries in the world. It's a country where millions of its own citizens have starved to death. Yet, this country is able to affect the actions of the United States, Japan, and China as a result of its ability to modify what is, in truth, outdated Soviet missile technology. As the *New York Times* indicated this morning, the Taepodong is little more than a longer-range version of the 1950s Soviet Scud missile.

But this illustrates the urgent need to address proliferation. Without real management of these materials and technology—much of it Russian in origin—it will become easier for third- and fourth-rate powers to drastically affect our own security decisions.

In truth, the greatest danger facing the American people today comes from nuclear weapons. Russian nuclear arms are the only threat that could kill every man, woman, and child in the United States. While I know your conference is focusing on the management of global nuclear material, the most potent challenge we will face in this regard comes from the nuclear legacy of the Soviet Union. It is on this issue that I would like to focus my remarks.

Without strong U.S. leadership, the Russian nuclear threat will increase in the coming decades. The truth is, Russian nuclear weapons present two distinct challenges. The immediate threat is from the possibility of an accidental or unauthorized launch. The second and more long-term threat is from the possibility of the proliferation of the weapons, their material, or their technology by dispirited, unpaid Russian military personnel.

Let me be clear. While I speak about the threat of Russian nuclear weapons, I do not believe a deliberate attack from Russia is likely or even plausible. However, as my former colleague Sam Nunn has noted, the threat from Russia does not come from its military might, but rather from its weakness.

To secure the post–Cold War peace, we must have the kind of U.S. leadership that won the Cold War. This leadership must come from the president, the Congress, and the American people.

To begin, the president must immediately take bold action to restart the arms control process. If we do not drastically reduce U.S. and Russian nuclear arsenals, the danger of their accidental use or their proliferation will increase exponentially. In order to facilitate further reductions it's time we concede for the foreseeable future that START II is dead. We can all make the case that the Russian Duma should have acted on START II, that ratification was more in their interest than in ours, or that the reason it failed was domestic Russian politics. All of this is true—but it does not change the need to move forward.

We have seen some recent signs of life in our arms control discussions with the Russians. But I do not believe we should take too much comfort in the announcement that the United States and Russia will accelerate negotiation of START III. Even if START III were negotiated and ratified in the near future, both sides would still possess over 2,000 nuclear warheads at the end of 2007. This would still be more weapons than we need to defend U.S. interests and more weapons than the Russians can safely manage. By encouraging Russia to keep a force larger than they can control, we increase the danger of both accidental launch and proliferation.

Simply put, I believe the START regime is too slow. The process takes too long because its safeguards were erected under a cloud of Cold War fears. Today, our relationship with Russia and our technology allow us to move more swiftly. The president must lead by finding new ways to work with Russia to quickly and dramatically reduce the number of nuclear weapons in a parallel, reciprocal, and verifiable manner.

We have a historical precedent to show that such leadership on the part of a U.S. president, based solely on an evaluation of our defense needs, can achieve the goal of reducing nuclear dangers. On September 27, 1991, then president George Bush announced a series of sweeping changes to our nuclear force posture.

On October 5—just one week later—President Mikhail Gorbachev responded with reciprocal reductions in the Soviet arsenal. The time is right for a similar initiative.

I recognize such deep reductions—while decreasing the chance of unauthorized or accidental launch—could actually increase the danger of material proliferation. Therefore, any such parallel reductions in our nuclear forces must include arrangements and a U.S. commitment to provide funding, to secure and manage the resultant nuclear material. We are fortunate that we will not begin from scratch on this problem. We can build upon one of the greatest acts of post–Cold War statesmanship: the Nunn–Lugar Cooperative Threat Reduction Program. To facilitate these dramatic reductions, we must look for ways to expand upon the success of this program, to enlist new international partners, and to work with the Russians to find new solutions to the problems of securing nuclear material. In this regard, I look forward to the ideas that will presented as a result of today's conference.

Leadership must also come from Congress. Too often we allow important national security matters to be held hostage to narrow interests. We fail to understand the broader definition of national security. In today's world, national security goes beyond tanks and fighter aircraft—it must also include strong intelligence capabilities and skillful diplomacy.

In this emerging nuclear era, this broader definition of national security will become increasingly important. Congress will need to seize the few opportunities presented to us to improve our capability to confront the nuclear threat and not fall prey to our tendency to demagogue and delay.

Ultimately, leadership must come from the American people.

Over the past year, as I have spoken to my constituents about nuclear weapons, it's apparent they no longer see a threat. It's certainly not like it was when I first ran for the Senate in 1988. At that time, people sensed the threat. It was real. They understood the danger of Soviet nuclear weapons. Today, my constituents believe the danger of nuclear weapons disappeared with the Soviet Union.

The reason I'm not asked about Russian nuclear weapons is that the American people have been lulled into a false sense of security. This comes as a consequence of their elected leaders repeatedly telling them the threat no longer exists. How many times have we heard that "for the first time in decades, American children go to bed without a single nuclear weapon targeted at them"? This statement may be factually correct but, as you well know, the journey from retargeting to launch is frighteningly short.

While we should not be alarmists, we must begin to do a better job of informing and educating the American people. This isn't just a task for people like myself and Pete Domenici who must face reelection. There is a role for the public policy community. Forums like this are a critical part of the education process. We must expand this dialogue to make it available to a wider section of the public.

Without the understanding of the American people, without their willingness to commit the resources necessary to address the problem, it will become more difficult to implement the kind of policy recommendations you are working on today. Just as it took an enormous effort to warn a tired nation at the end of World War II of the dangers posed by communism, we must do a better job of explaining to the American people the costs and consequences of not following through and securing nuclear materials.

It was U.S. leadership that won the Cold War—leadership not only in the international community but also domestically. In this new era, it will take the same formula and abundance of leadership to secure us from the nuclear legacy of the Cold War.

CHAPTER 7

A View from the Outside

Remarks by James R. Schlesinger

I shall start with a bit of reminiscing. Some 36 years ago, I was the project leader of the RAND Corporation study on nuclear proliferation. At that time, the Kennedy administration was determined to stop Charles de Gaulle from acquiring nuclear weapons. We had some rather ingenious policies like the proposed multilateral force (MLF) that later was abandoned with some embarrassment. The administration's way of dissuading the French was to inform them that they were so technologically backward that they couldn't produce nuclear weapons anyhow—something which they found rather offensive. That just infuriated—and spurred—de Gaulle. I personally informed the Pentagon that, if the United States wanted to stop de Gaulle, the only way would be to bomb Pierrelatte, the French gaseous diffusion plant. In brief, that particular policy in nonproliferation was doomed to failure.

It is interesting that in the 1960s there was every expectation that the number of nuclear powers would expand rapidly. President John Kennedy himself mentioned 15 to 20 nuclear nations. C. P. Snow, an author of that period, said, "within a decade some of those weapons will be going off." Thus we must bear in mind that we have done far better on preventing the spread of nuclear weapons than anybody expected back in that period—partly as a result of Cold War discipline. That Cold War discipline is now gone, and a question before us is whether or not we might be returning to the expectations of that earlier period.

To turn to the current scene, I shall begin by telling you about an old farmer in County Armagh, who was too old to dig up his garden and plant his potatoes for the year. So he wrote to his son, who was in prison for IRA activities, what his perplexity was. The son wrote back immediately: "Don't dig in the garden. That's where I hid the weapons." The next morning at 6:00 A.M., 12 British police officers arrived with shovels. They proceeded to dig in the garden for several hours but were disappointed at the end. The old man wrote to his son in prison describing this sequel, and he got this reply: "Now you can plant your potatoes."

I think that there's a metaphor here that tells us something about the current scene. With the end of the Cold War, it is clearly an act of madness on the part of any nation militarily to challenge the United States—because of our overwhelming conventional superiority. It is this that drives other nations toward asymmetry, that is, a way of effectively dealing with the United States without directly challenging it militarily, at least in conventional terms. That asymmetry involves not only the possibility of nuclear weapons, but also chemical weapons, biological weapons, and cyberwarfare, all of which could potentially be used to deter American activities, either military or political.

Another consequence of the end of the Cold War is that the United States has taken to strutting around like the cock o' the walk—in the course of which we are serially irritating everybody else. (That, by the way, is not the way to elicit cooperation on matters of weapons of mass destruction.) Because of U.S. conventional superiority, the fact that the United States is driving as hard as it is on questions of chemical, biological, and nuclear warfare is regarded with some suspicion. Naturally, other nations conclude the United States wants others to play by the new rules, which involve perennial American conventional superiority. That is a suspicion that we shall have to counter.

There are repercussions also of our new position and our actions. As a result of the Gulf War, the Indian chief of staff (COS) observed that the lesson of the Gulf War—you will remember lessons of Vietnam; we are now engaged in lessons of Kosovo—but the lesson of the Gulf War, said the Indian COS, is that no nation should challenge the United States unless it has nuclear weapons. That was some years before the recent nuclear test by India. In the wake of Kosovo, [former prime minister Yevgeniy] Primakov said publicly—and [former prime minister Viktor] Chernomyrdin privately—that the consequence of our actions in Kosovo was to drive other nations toward weapons of mass destruction, notably nuclear weapons. And now the North Koreans have said much the same thing. So our position of military preponderance in the conventional area has repercussions elsewhere. It is something that we need to have clearly in mind.

With respect to the question of the spread of nuclear weapons, let us deal with the major players. First, Russia. You have dealt with the problem of Russia's nuclear complex extensively this morning. Russia is so preeminent a problem in this area that some would wish that the disciplines of the Communist Party were once again applicable throughout the Russian defense establishment—instead of what is a very weak system of command and control. Russia, of course, could return someday as a considerable military power, but that is years away. In the interim, we have had observations from senior figures in Russia urging that, after Kosovo, Russia should now turn to the development of new tactical nuclear weapons. One will note that the Russians have moved in the direction of United States and NATO military strategy of the extended Cold War period when the United States and its European allies felt that they were at a severe disadvantage in terms of conventional strength vis-à-vis the Warsaw Pact and that the way to deal with that was to respond early on, if necessary, with nuclear weapons. The Russians have moved into that position and probably will do so increasingly. That is the lesson of Kosovo for them.

Another aspect of Kosovo is the discussion earlier by Roger Hagengruber about the political uncertainties in Russia. I think that it is clear that next December we are going to have a Duma that is more recalcitrant than the existing Duma. The consequences of Kosovo have been largely a wipeout of those who are our friends in Russia—those who have been closest to the notion of reform and liberalization. We are going to see a Duma that is more nationalist and more Communist than the present one. That raises the question of what we should do under circumstances less benign than those postulated this morning.

Second problem. You have no doubt had the pleasure of reading stories about the Cox Report and of the Chinese "theft of our nuclear secrets." I point out, in passing, that under international law, espionage is not illegal. Although it may be illegal under American law, under international law it is regarded as an accepted state activity. And I add that if you spread your goodies out on the table, do not be too surprised if somebody comes along to pick them up. The Chinese have lately talked about "a restoration of the true spirit of partnership." This follows our anger with them over Kosovo and espionage and their anger with us over certain unfortunate events such as the bombing of their embassy in Belgrade. The Chinese conception of the true spirit of partnership is, in simple English, tit for tat. One notices that in the wake of those developments in Belgrade, the Chinese have lately been accused of shipping accelerometers and gyroscopes to North Korea, which would help with the North Korean missile program. This is quite plausible in light of Chinese attitudes toward fair dealing.

State assistance has been quite commonplace with respect to weapons of mass destruction. The Chinese undoubtedly helped the Pakistanis to design their nuclear weapons. Russia is helping Iran with regard to at least the Bushehr reactor and with its missile program.

Now, while I am mentioning these activities of other nations, I should not refrain from paying tribute to what, no doubt, has been the greatest proliferation source in the world—to wit, the United States of America. While we talk about preventing nuclear proliferation, we need to examine our own conduct throughout history. Back in 1963, when I was the project leader at RAND, to learn about the weapons design (having gotten top secret and Q clearances), I was taken into a little room and design concepts were whispered to me—whispered things that to my astonishment now appear daily in the newspapers. They have now been declassified—as an example of openness, I suppose. They are an immense help to anybody who wishes to follow in our path.

We must remember also that we were the original weapons state and that we have given assistance to others to join us in that category, some of it deliberate, as with the United Kingdom and with France; some of it indirect, as with Israel. Roger Hagengruber: Tom Graham asked you about three countries—Pakistan, India, and Israel. While you batted .667, you failed to respond completely. You did not reply with regard to Israel. I don't want to put you on the spot, but the prospect of the United States applying pressure to Israel with regard to its nuclear program strikes me as questionable at best. Intentionally or not, the United States gave indirect help to Israel. It was not policy—but it was reality.

And, under the heading of reality, we must acknowledge that we gave substantial help to Saddam Hussein—in declassifying the technology for magnetic separation on the rather quaint premise that that technology was not cost-effective for us.

Finally, North Korea. As a result of the Framework Agreement reached in 1994, instead of a 60 megawatt reactor we are going to provide 6,000 megawatts of thermal capacity in North Korea if KEDO (the Korean Peninsula Energy Development Organization) achieves its objective. Moreover, no doubt, as part of our policy of openness, the DOE has advertised that one can make a pretty dandy good weapon just with reactor-grade plutonium. Of course, that again underscores the question of whether openness has not gone too far.

Roger Hagengruber indicated in his remarks that there is some question about whether transparency can be differentiated from spying. This policy of openness is one in which we have gone much further than we should have. Some of it reflects the hopes and the illusions of the United States in the post–Cold War world—illusions that should be shattered by such developments as the Indian and Pakistani nuclear tests. If you will recall, the present government of India, at least the transitional government, declared the last time that it was in office that it was going to test nuclear weapons. That became part of its election platform. When it ran again two years ago, it said: We are going to test nuclear weapons. Of course, we thought we knew better what the Indians would do than the Indians did themselves, so we brushed off the notion that they might proceed to a nuclear test. In fact, the alertness of our intelligence establishment on that issue was not at an all-time peak.

We also have this illusion, by the way, that all knowledge is here in the United States, so that any other country that develops a military technology must have stolen it from us. Subsequent to the Indian test, there was a hullabaloo here about our having sold an American-made supercomputer to the Indian Institute of Science. There was an interview with the head of the computing arm of that institute in which he discussed the Indian supercomputer. He observed that it was just a collection of Sun microprocessors (available on the global market)—and that others had acquired that kind of capability by putting together IBM microprocessors. He went on to say, "You know, it's 50 years since the Americans detonated their first weapon and many of us have had an opportunity to study this for an extended period of time. On my desk I have available 10,000 times computing power as the Americans had in 1945." And he threw in, perhaps unkindly, "By the way, have you ever looked at the information provided on the Web sites of the nuclear laboratories of the United States?" I do not think that our delivery of the supercomputer was the basis for the Indian detonation. One should note that our explanation for the sale was that it was delivered to a civilian scientific site and therefore was legitimate. The British government, by the way, had identified the Indian Institute of Science as a dual-purpose activity.

Let me say a few words more about transparency. Transparency usually means other people should be transparent. Transparency doesn't necessarily apply to us. There is, of course, a good deal of legitimacy in that point of view, but it also descends to the level of "You are opaque, we are transparent." Therein lies, I believe, the problem of seriously monitoring large programs. I do not have time to go into monitoring the nuclear weapons dismantling program in Russia. But, for security reasons, we are reluctant to expose our own weapons dismantling to the Russians. For some strange reason, they are unprepared to disclose their dismantling of weapons to us—unless we are prepared to reciprocate. That may seem unfair to some of us, but that is the position that the Russians have been taking. That means we must estimate dismantling largely through guesswork—and through accepting what the Russians tell us.

Some of you may have had the opportunity of reading Ken Alibek's *Biohazard*. For those of you interested in either monitoring weapons programs or monitoring civilian programs, I strongly commend that book to you. This is with regard to the steady Soviet violation and expanding violation of their signature on the Biological Weapons Convention of 1972. It underscores the tremendous challenge of verification.

Let me sum up. What we face here—and I do not want to overstate it—is a very formidable challenge. The disciplines of the Cold War are now gone; the amount of information out there is larger; and the temptation to reach for nuclear weapons for a variety of reasons, including the desire for an asymmetrical capability for deterring the United States, has grown.

As I indicated earlier, we have been remarkably successful compared with our expectations of the 1960s. But it is important for us to recognize the limitations of what we can do. It is important for us to avoid hysterical reactions when some unfavorable event takes place. What we will probably see over the decades ahead will be a slow, slow retreat on the issue of nuclear spread. The disappearance of the Cold War disciplines is a serious problem. The reaction to our overwhelming conventional military superiority will be a difficult problem. Yet, one happy reality is that probably only nations will reach for nuclear weapons. The likelihood of these trickling down, as it were, to subnational groups, to terrorists groups, is, I think, much less than is frequently stated. If a self-respecting terrorist group wants to reach for an asymmetrical capability vis-à-vis the United States, the relatively easy way to do it is to pursue biological weapons rather than nuclear weapons. The latter path is fraught with many more difficulties.

Nonetheless, the rules of the game have changed since the end of the Cold War. We face a formidable challenge. We shall have to do the best we can, but we should not hope for total success.

CHAPTER 8

A View from the Administration

Remarks by Under Secretary of Energy Ernest J. Moniz

It is a pleasure to come here and speak before this workshop on global nuclear materials management. We are, as you all are quite aware, indeed faced with profound challenges in dealing with the nuclear material legacy of the Cold War and also with new opportunities and unmet challenges for the emerging post–Cold War period. CSIS is certainly to be commended for pursuing this project on nuclear materials and for emphasizing the need for action now. I certainly recognize the important role that organizations such as this can play in moving this forward on these agendas. In that regard, certainly I and all of my colleagues at DOE look forward to working with this organization and all of you here.

The CSIS project suggests an ambitious goal, certainly, for the United States, and I quote, "to lead a world where all materials are safe, secure, accounted for from cradle to grave with transparency." As has been stated already, Russia of course presents a special challenge because of the combined factors of enormous nuclear infrastructure together with economic and political uncertainty. And so today I will certainly focus my remarks on our programs with Russia. Others should note that Rose Gottemoeller, whom I think many of you certainly know and who heads the DOE nonproliferation program, is here as well; and she has many of the real on-the-ground responsibilities for getting these programs implemented.

Let me start by trying to frame my remarks before coming back later to some more specific programmatic comments. On the one hand, we have made considerable progress in the last several years in our dealings with Russia in addressing what's called military nuclear management issues. This has been the result of our work together to deal with the legacy of the Cold War. Our scientists have worked very closely with their scientists building on a shared sense of responsibility in addressing this military legacy and frankly on a pride in their contributions over decades to their and, of course, our national security programs. That focus is evident in the progress I think that we have been able to achieve, admittedly with lots left to accomplish—in policies and programs like arms control, delivery and warhead systems, MPC&A, disposition of HEU, progress toward plutonium disposition, and assisting the transition of scientists and engineers in Russia into nonweapons work.

On the other hand, I would say we have done relatively little in cooperation on civilian nuclear power. Of course we have differences in starting point with Russia about, for example, the potential proliferation risks in the nuclear fuel cycle; and this difference manifests itself in strong disagreements over Russia's civilian commerce with third countries such as Iran. It makes uncertain our plans for nuclear cooperation with Russia on this front, particularly since the political impetus to collaborate on the civilian side is clearly weaker, a fact that can be witnessed in the budget discussions, for example, that go on in terms of support of the various programs. So this diminishes our opportunity to use civilian cooperation to leverage dealing with the Cold War legacy. But those opportunities, at least in my view, are quite real since, as we all know (yet need to repeat as often as possible), the key nonproliferation issues for both military and civilian sides in the end revolve around weapons-usable nuclear materials and the knowledge and technology needed to produce or acquire them.

Twenty years ago there was an intense debate throughout the world, certainly in this country, on the proliferation aspects of the commercial fuel cycle. The United States then chose a different path from that chosen by several other industrialized states, rejecting a fuel cycle involving reprocessing. Others' primary concern was energy security, ours was nonproliferation. The result was a divergence with results that are even clearer today than before. For example, today more than 200 tons of civilian-separated plutonium are in storage around the world and tens of tons of additional material are being processed every year. Some states that favor reprocessing have accumulated more plutonium than they can use. And the economics of plutonium utilization have proved unfavorable, at least up to now. While some states that once favored this option are turning away from it or scaling back in their programs, we still see this concern expanding into the future, absent changes in the fuel cycle. We also see nuclear power expanding in some parts of the world, with spent fuel accumulating and introducing new waste management challenges and additional nonproliferation risks. In those 20 years we have also seen of course a couple of significant nuclear accidents. We've seen economic challenges to nuclear power and limited progress on waste management.

Having said all that, however, we are in my judgment now entering a critical period in terms of the civilian sector due to the accumulated and real concerns over waste management, safety, and economics. Again, the waste management issue has not been resolved to the public's satisfaction although much progress is being made, certainly on the scientific basis for long-term geological disposal. No nuclear power plants, no new plants, have been ordered in the United States for many years. Deregulation of electricity markets worldwide will only exacerbate anxieties about major capital investments in new nuclear facilities, particularly in the absence of a path forward on nuclear waste. At the same time, nuclear power has clear environmental benefits for regional issues of acid rain and global issues such as climate change; and this has served to reactivate the nuclear power dialogue in this country and elsewhere.

No one can presume to know how this balance will work out between the increased energy demands and environmental benefits of nuclear power versus the unresolved issues of nuclear waste, for example. But I do believe that the United States has an obligation to research options for the future and, given—at least at the DOE—our combined responsibility for energy and security, to develop options that minimize proliferation concerns. Clearly this can only make sense in the international context, as we learned 20 years ago with the reprocessing decision. Consequently this is a period of opportunity for the United States to rethink the international path forward in the civilian sector and perhaps to reengage the fuel cycle discussions in the context of our nonproliferation interests. I believe, as I will now turn back to the Russia program specifically, that this linkage between the military and civilian sides in the nonproliferation context is a very important one for us to move forward.

So the problems of energy activities are of course entwined in many of the issues I just raised, and I would briefly now touch on five questions and issues that this policy forum is organized around: nuclear materials security and purchase, commercializing or transforming excess defense infrastructure, spent fuel issues, transparency, and status and future of the U.S. nuclear infrastructure.

The first question is, What are we doing to secure and purchase nuclear material from the former Soviet Union and what more could be done? Obviously the key materials of concern are HEU and plutonium, military or reactor grade. I will not go through a detailed listing of all our programs, but let me just go over a few of them.

First, as you know, the MPC&A program, the materials protection, control, and accounting program, certainly has in the last several years secured a very significant amount of weapons-usable material in Russia and other parts of the former Soviet Union. When this program started in 1994 at one site in Russia, the working premise in the Congress was that we were going to complete the task in Russia within five to seven years and basically depart. Now, with more than 50 sites in the former Soviet Union, we have learned a lot in the intervening period. The National Academy recently recommended the program in fact be extended into the future, and in many ways it grew in an effort to further consolidate and maintain these critical materials. Clearly the Russian economy has been a major factor in this changed perspective, particularly in regard to the issue of trying to institutionalize and sustain in fact even the already implemented arrangements. Another phase clearly involves consolidation, a task that we need to address in this country and a task in Russia that in fact has the support of leadership at MINATOM.

Let me just digress slightly and make a broader comment on this question of the Russian economy. Fundamentally the lack of a functioning, cash-based economy obviously presents many challenges in our nonproliferation work with Russia. We just mentioned the implications it's had for extending the time horizon for our assistance in collaboration in MPC&A. It obviously, as we will discuss shortly, affects chances for success and things like transitioning in the defense complex by creating business opportunities. But it also affects our nonproliferation aims in other ways, and this is where also there is a linkage to the civilian sector. For example, the absence of a growing economy that demands major new energy sources limits the possibilities for financing new reactors that could help with plutonium disposition or even conventional energy sources that might be a path to eliminate plutonium production. So, again, military and civilian sides are strongly coupled and, as we wait for the Russian economy to get into gear, many of our programs are challenged.

Second, in terms of purchase and disposition, as I think you know the 1993 HEU purchase agreement is basically back on track. This is a vital underpinning of our nonproliferation cooperation, with Russia—with highly enriched uranium from weapons being blended down into reactor fuel—receiving ultimately many billions of dollars in revenue. There was a problem with this agreement largely foundering on the so-called natural uranium component. With the help of Congress through an allocation of $325 million for us to purchase the backlog of natural uranium, that agreement is basically functioning. We have in fact largely completed the purchase of the natural uranium, and the LEU from Russian weapons in fact has been flowing back to this country with the 1998 requirements having been met. This is an agreement based upon market activity; and once again we come to the economic rail where the health of the market, in this case for natural and low-enriched uranium, inevitably complicates these activities. We can expect that to recur in a certain sense as one has to address the terms of the agreements over the long term as the markets go up and down.

With regard to HEU, I will just note that of course the possibility of what the task force called HEU-II certainly hasn't escaped our attention. It's certainly a very interesting possibility. There are clear national security benefits from providing incentives to Russian research institutions, for example, to close out caches of HEU or similar fuels, perhaps convert their research reactors. But any such initiative clearly again runs into this market issue: that is, we have to address these kinds of initiatives so as not to affect market conditions for the HEU purchase agreement and, in fact, broader market conditions. And here the uncertain prospects of nuclear power have serious impacts on the uranium market so that once again the civilian realities come and impinge upon some of our nonproliferation goals by limiting our flexibility. There are also impacts that have to be viewed in terms of our own infrastructure transition in the United States, particularly that associated with uranium enrichment.

Third, I should also comment on plutonium disposition, where again there is progress but in this case certainly a very long way to go. This is the issue that the National Academy of course termed "a clear and present danger." We are trying to negotiate, under the leadership of John Holum at the Department of State, an agreement to mutually dispose of the first 50 tons at least of military plutonium. Several complications exist: in contrast to the HEU deal with plutonium disposition, we are dealing now with a proposition that cannot be based upon market incentives at this time. There are simply costs to be borne and rather substantial ones at that. We are also impeded by the ability in Russia to burn MOX with current reactors. MOX fuel can be consumed at a maximum rate of about two tons per year, which of course would present a rather long timeline to meet this program. Furthermore, just to note—and again we keep going back to the civilian connection—that plutonium is still being separated in Russia, in fact at a rate comparable to that at which they can burn military plutonium, although there is no credible economic use for it in the foreseeable future.

Progress I think has come from the fact that there is a strong engagement with Russia in terms of trying to reach this first-phase agreement. Second, on the technology side, the discussions have moved to the production of a real technology road map that is aimed at disposition as opposed to a program that is much more broadly based in terms of research.

The second question that your workshop has raised is, What are the possibilities for commercializing or transforming excess defense infrastructure, particularly in Russia, and what more can be done? This issue of infrastructure when all is said and done, at least from our perspective at DOE, is ultimately about people. And it's difficult. First, our IPP program, Initiatives for Proliferation Prevention, illustrates the challenges and the importance of persistence. The program has had important successes. To date it's engaged more than 6,000 former Soviet weapons scientists in Russia, Belarus, Kazakhstan, and Ukraine, with 3,000 currently active on more than 200 projects. Nearly 80 of these projects are commercial partnerships to which industry has contributed more than $60 million, with nearly 20 percent of these projects expected to reach commercial status this year.

Commercialization is clearly difficult in Russia, as we alluded to earlier. It isn't that easy in the United States, for that matter, for small businesses, but the difficult economic conditions in Russia remind us that the pressures on individuals of concern may only increase and that commercialization is a means to an end, not the sole measure of interest. It's much more difficult to provide a matrix of success for these programs in comparison with something like the HEU or MPC&A programs, and this leads to difficulties in sustaining support. But, as Einstein said, "Not everything that counts can be counted, and not everything that be counted, counts." Working on these people issues, in my view, certainly really counts. We are tightening up management of the programs certainly to maintain support as best we can. We need your advice and support. Again, we remain committed to these people programs as being core to our nonproliferation work.

Second, last year we also began the Nuclear Cities Initiative (NCI), which is focusing first on 3 of the 10 nuclear cities in Russia, helping them transition to a smaller nuclear complex. Again, this is a very important and key nonproliferation goal and one that has a clear coupling to future arms control activities. But, again, we are dealing with human capital here. We have had some success with establishing small nondefense enterprises and business centers to encourage more. The overall initiative thus includes investments in preventing the most immediate and worst nonproliferation concerns from materializing, even as we seek to establish stable outcomes in the U.S. interest.

Third, to look further ahead, an important possibility emerging out of the NCI is joint work to accelerate planned closures of important facilities. Such facilities include, for example, plants involved in the serial production of nuclear weapons. Several of these cities, as the CSIS task force observes, have more weapons-usable material alone than the combined arsenals of Britain, France, and China. Our objective is not to intercede into such facilities but, rather, a mutual effort to relieve the economic burden of a too-large nuclear complex. I would observe that at DOE we have in the last five years come down by about 50,000 people, almost one-third of the total DOE federal and contractor workforce. This has not been easy for us. Working with the communities has not been easy, and yet that's in a country with obviously a rather robust economy. In any case, we are working closely with Russia. Russia's publicly stated decision to close down the Avangard and Penza 19 plants holds the promise of an opportunity, funding permitting, to assist in this "right sizing" of their nuclear complex, just as we have and continue to adapt our own.

In this context, as well, an interesting idea that is out there in discussion is that of complementing, not replacing but complementing, the work in terms of long-term development of sustainable business opportunities in these cities and to consider—and I stress to consider—the possibility of getting involved more in terms of providing direct research opportunities for issues on which we need deliverables. There are many areas. Environmental remediation is a good example. This can provide some short-term benefits, both in terms of a goal for the people problem in Russia, but also in terms of some of our issues where we can draw upon the large Russian creative, talented establishment. It's also clear this has considerable issues, political issues, for example, that would need to be addressed; and once again your advice and support would be quite welcome.

There are other major issues with infrastructure, such as submarine dismantlement, but I think I will forgo that for the moment and move on. The third question is the issue involving spent fuel and what more could be done, particularly with Russia.

First, a reminder that, as you all know, disposal in a deep geological repository is a fundamental component of the U.S. nuclear materials stewardship responsibilities. Ever since the atoms-for-peace program, permanent isolation of nuclear waste has been a significant part of our comprehensive strategy. As the U.S. and international nonproliferation objectives and strategies have evolved, the repository program has addressed the challenge to ensure safe and proliferation resistant disposal capability. A geological repository provides safe isolation and is an end state that is needed to meet long-term national security goals no matter what spent fuel management approach is adopted. We are developing the Yucca Mountain site. This has been a very long process, and we all know the difficulties in reaching agreement on geological disposal. Earlier this year we did open the WIPP (Waste Isolation Pilot Plant) site in New Mexico for low-level waste after a very long period. For Yucca Mountain, I feel that we have developed in the last years by far the most extensive scientific characterization, scientific study in engineering development related to geological disposal of high-level waste. There will still be challenges as we enter the suitability determination and licensing stage in the next several years. Also, at the end of October, the secretary of energy will be hosting an international conference of policymakers addressing the questions of geological disposal in the international context. There is clearly a growing need to address long-term spent fuel management on a global scale as the conference will begin to do.

Proposals that involve the return of spent fuel to Russia certainly are interesting and intriguing although it is clear there are a number of very difficult issues that have yet to be engaged. This is the kind of thinking that groups such as yours can certainly help to shape. The United States, for example, would certainly have its own views on requirements to support the storage in Russia of third-country U.S.-origin fuel and concerns about ultimate disposition and management of the material.

The fourth question that you've raised is about transparency concerning secure storage and legitimate use of nuclear materials internationally. DOE is responsible for safe and secure management of its own materials in ensuring transparency in accordance with our international agreements. The department's nuclear materials management obligations are certainly fundamental, compelling, and enduring, and we need to support U.S. leadership in this area. The department has a special responsibility on matters of international nuclear materials transparency, such as at the IAEA, and we must clearly set an example for responsible management here at home. Some of the techniques employed in DOE–MINATOM transparency efforts—remote monitoring, for example, involving sensors, computers, and real-time data—have already been examined in a broader context by the IAEA and may hold some promise of broader cost-effective applications. The structure, however, that exists as we look generations ahead is not yet fully adapted to the future nuclear materials stewardship mission. The word stewardship is taken in the sense that as we have in the department an ongoing stockpile stewardship mission, we have certainly an increasing materials stewardship mission, as well as a responsibility to future generations. We are trying to develop a more comprehensive strategy that will enable the department to meet these legacy obligations responsibly; and Russia, of course, has many of the same challenges.

There are other elements of transparency in terms of the military side that I will forgo for the moment. The fifth question is, What is the status and future of U.S. domestic nuclear infrastructure and R&D? First, I think one of the important issues is that although the eventual outcome in terms of the extent of nuclear power utilization in this country remains unclear, as we look at a decade and more ahead, the DOE and the secretary feel very strongly that we have a responsibility to explore energy options and to do so not only for our potential energy needs but also to do so in the context of maintaining leadership in nonproliferation issues.

In that context, the secretary has chartered a new advisory committee, the Nuclear Energy Research Advisory Committee (NERAC), that will help assess the department's R&D facility infrastructure for the next 20 years. And this year with strong congressional support, we have started a new R&D program called NERI, the Nuclear Energy Research Initiative, that is a peer-reviewed, competitive program that focuses again on the long-term fuel cycle questions, using such criteria as proliferation resistance, waste minimization, and safety, for example. So these initiatives are contributions to sustaining a nuclear infrastructure. But we all know that there are also problems and challenges here: the nuclear power sector has not flourished in this country certainly in the last years and, as the Cold War ended, we see challenges in terms of training new people, bringing in young scientists and engineers. Recently in the context of our stockpile stewardship program, Chile's commission emphasized a key challenge in stockpile stewardship in the years ahead is in fact people, and the same could be said about our material and nuclear energy stewardship responsibilities. This is an issue I think we do need to move forward on. We have taken small steps in terms of the educational arena, but again we will be looking forward to your recommendations and to your further suggestions.

While it's often frustrating to see what appears to be slow progress, I think if one looks back over the last five years, one will find that in fact a lot of progress has been made in terms of programs with Russia. But the message I do want to repeat is that conceptually there is a broad buy-in to supporting our collaborative programs with clear goals, particularly when attached to the issue of weapons and military-origin materials. We, in my view, frankly don't have that kind of broad buy-in to the programs that are not amenable to simple metrics and that involve civilian nuclear power sites. Yet as we have seen, there are so many linkages between the two, I think (a) it is very hard to separate and (b) exercising that linkage would give us a lot more leverage in many of our nonproliferation programs. This is an opportunity for us now, and the question really is how to engage in that fuel cycle discussion in a way that we can reemphasize nonproliferation goals.

CHAPTER 9

Looking Ahead

Remarks by William C. Potter

Introduction

It is a great honor to address this august gathering on a subject that arguably is the number one national security priority for the United States as we approach the new millennium. It also is a daunting assignment as the closing speaker at such an intellectually rich conference to say something for 10 minutes that is both new and relevant. In approaching this task, I was inspired by the wisdom of George Bernard Shaw, who was once approached by an acquaintance who was appalled at the prospect of telling a scientific gathering everything he knew in only three minutes. Shaw's advice was, Speak slowly.

Finding myself in agreement with Shaw as well as most of the recommendations contained in the draft I read of the report of the senior policy panel, I will confine my remarks to three vital issues about which more might have been said regarding the need for U.S. leadership:

- The fragile state of health of the Nuclear Non-Proliferation Treaty (NPT);
- The growing potential for a major accident in the civilian nuclear power sector in the former Soviet Union; and
- The need to train a new generation of nonproliferation specialists in the United States and abroad.

A Fragile NPT

My first point is that although the indefinite extension of the NPT in May 1995 was a major accomplishment, the treaty and the broader nuclear nonproliferation regime have experienced a number of major setbacks in the post-1995 period that, if not reversed, threaten the long-term viability of both the nonproliferation and peaceful-nuclear-use provisions of the NPT. Some of the setbacks are well known and include the 1998 nuclear tests in South Asia, Iraq's defiance of the United Nations Special Commission on Iraq (UNSCOM), North Korean nuclear and missile brinkmanship, and the growing difficulty of safeguarding Russia's vast arsenal of nuclear material and technical know-how. Less well known, but also detrimental to the NPT's health have been the muted and ineffectual international response to the South Asian tests, the unwillingness of many NPT parties to take seriously the three decisions in 1995 linked to the indefinite extension of the NPT (namely, the strengthened review process, the principles and objectives for nuclear nonproliferation and disarmament, and the resolution on the Middle East), and the erosion of

U.S.-Russian cooperation for nonproliferation. Without being overly dramatic, let me suggest that these three developments already have led to the reopening of debates in a number of countries about the value of the NPT for their national security and that failure to address these challenges in a timely and concerted fashion could lead to a number of NPT defections, including that of Egypt, Syria, Yugoslavia, and—under certain circumstances—even Ukraine, South Korea, and Japan. What I find most troubling is that many of my good friends in government—in Washington, but also in Moscow, London, and Paris—don't even have these possibilities on their radar screens. Indeed, I can envisage realistically a situation next May on the day after the end of the 2000 NPT review conference when the *Washington Post* contains four news headlines:

- Headline One: Russia, China, and the United States resume nuclear testing.
- Headline Two: Yugoslavia defects from the NPT and revives its nuclear weapons program, dormant since 1987.
- Headline Three: The new Ukrainian government follows the instructions of the Rada and repudiates its NPT commitments.
- Headline Four: World braces for fallout from accident at the Kursk nuclear power station.

Buried on page 27 of the paper is a one-sentence news brief that quotes the chairman of the U.S. delegation to the 2000 review conference to the effect that because the conference produced a final document, the review conference was a success and the NPT is in good shape. I fear that the only unrealistic part of my hypothetical news stories is that the *Post* would not include the news brief on page 27 because it rarely covers NPT review conferences.

Nuclear Safety

The fourth headline I mentioned, regarding an accident at a Russian nuclear power site, I hope will never materialize. Nevertheless, just as greater U.S. leadership is essential if we are to meet the challenge of safeguarding Russia's vast stocks of nuclear material, so too must there be more leadership if we are to cope effectively with the continuing threats posed by unsafe nuclear power practices in the former Soviet Union. These risks, although moderated at certain facilities thanks to Western and Japanese assistance, remain acute at many plants in Russia, Armenia, and Ukraine. Severe resource constraints, weak regulatory bodies, a die-hard *Reaktory Bolshoi Moshchnosti Kanalyne* reactor lobby, and an underdeveloped safety culture combine to vitiate many of the post-Chernobyl safety improvements. So strapped for cash is the nuclear industry in the post-Soviet states that many facilities have virtually stopped reactor safety maintenance and repairs. At least partially as a result, the number of reactor malfunctions last year reportedly increased by 20 percent in Ukraine, while in Russia fuel shortages and delays in repairs have forced nuclear power plants today to operate at one-third of their output of two years ago. They simply lack the money to pay employees, perform maintenance and repairs, and even buy fuel. Employees, for their part, are discontented, distracted, and all too often disinclined to follow modern safety practices, tendencies that impoverished and understaffed nuclear regulatory bodies are ill equipped to correct. On top of these risks is the growing threat of terrorism directed at post-Soviet civilian nuclear power research facilities—many of which are very vulnerable to sabotage and attack.

My purpose in raising these dangers is not to suggest that another Chernobyl is imminent—although it shouldn't be ruled out—or that the problems of nuclear safety are amenable to quick fixes. Any study that seeks to confront squarely the issue of post-Soviet nuclear materials management, however, must be attentive to the very real risks of safety and security in the civilian nuclear power sector, especially at a time when the nuclear industry in all of the former Soviet states is experiencing tremendous economic strains.

Training the Next Generation of Nonproliferation Specialists

The last issue I would like to discuss pertains to rebuilding the foundation for U.S. nonproliferation leadership. I am very pleased that the draft report of the senior policy panel includes a section on this topic because we too often take for granted the existence of a self-perpetuating cadre of nonproliferation specialists who are competent on both technical and policy issues. In fact, the opportunities for systematic study of nonproliferation are very limited. Few, if any, high schools have curricula that expose students to issues of weapons proliferation or strategies for their control, and the possibility for university training is not much better. To the best of my knowledge, only one graduate school in the United States offers a formal concentration in nonproliferation studies.

The U.S. government to date has failed to appreciate how education might be used as a nonproliferation tool. As a consequence, there is a tremendous gap between the bipartisan, high-level government statements that proliferation of weapons of mass destruction is the paramount threat to U.S. national security and the paltry amount of money allocated to train the next generation of nonproliferation specialists.

One possible approach to correct this situation would be to craft a National Nonproliferation Act, perhaps modeled after the National Defense Education Act or the National Security Education Act. Such legislation, funded at a very modest level, could provide fellowships to graduate students for advanced, multidisciplinary training in nonproliferation at the university of their choice.[1] Another approach worth exploring, and one noted in the Deutch–Specter commission report, is the use of an ROTC-type program in nonproliferation studies in which the government would offer financial support to individuals pursuing undergraduate or graduate training in the field of nonproliferation in exchange for a commitment to work in the government after graduation.

---

1. The National Security Education Program's funding mechanism was a one-time appropriation of $150 million in the form of a trust account. The program has operated on the annual interest from that account, amounting to approximately $7.5 million for programs and $1.5 million for overhead annually.

Conclusion

We have come a long way in dealing with the issue of nuclear materials management since the idea for the Nunn–Lugar Cooperative Threat Reduction Program was born in the U.S. Senate in the fall of 1991. The proliferation challenges we face, unfortunately, also have grown dramatically. It therefore is imperative that the United States, collectively with Russia and the international community, combat proliferation with a will and budget commensurate with the threat. Otherwise I fear our efforts may resemble the story I was told by one of our recent graduates now working at the U.S. embassy in Moscow. He reports that last year there were three-man government-employed teams that went out and planted trees in the city. One day our embassy friend saw two men out along the road. One was digging holes and the second was filling them. Curiosity got the better of the embassy observer, and he asked the two men what they were doing. Why was one digging holes and the other immediately filling them? The Russians replied that they were part of a three-man tree planting team. Member number two was sick that day. His job was to plant the tree. Numbers one and three still had to work to get paid and, thus, continued on without the crucial third party.

We cannot go about our business in a routine way like the Russian tree planters. The United States must show up for work in order to plant and nourish the seeds necessary for a global nonproliferation culture to take root and flourish. Given our enormous budget surplus, let's invest at least a small portion of those monies in tackling the difficult proliferation challenges so eloquently detailed in the senior policy panel's report.

Conference Agenda

Phase II
The Emerging Nuclear Era—Policy Recommendations

July 22, 1999

8:30 A.M. **Registration and Continental Breakfast**

9:00 A.M. **Welcome**

Richard Fairbanks
President and CEO, CSIS

Opening Remarks

Sam Nunn
Former U.S. Senator
Chair, Board of Trustees, CSIS
Chair, GNMM Senior Policy Panel

9:15 A.M. **Funding Nuclear Security**

Should the United States buy all the weapons-usable nuclear material from Russia to ensure it remains safe and does not fall into the hands of rogue states or terrorist groups? If so, what would it take and what would it cost?

Matthew Bunn
Assistant Director of the Science, Technology, and Public Policy Program, Belfer Center for Science and International Affairs, John F. Kennedy School of Government, Harvard University
Vice Chair, GNMM Task Force I

10:00 A.M. **An International Spent Fuel Facility and the Russian Nuclear Complex**

What role could an international spent fuel facility play in helping the Russian nuclear complex transition? What benefits, if any, would ensue for Japan, Korea, Taiwan, and other potential participating states in the region? Is such a facility financially feasible?

Atsuyuki Suzuki
Professor of Nuclear Engineering, Department of Quantum Engineering and Systems Science, University of Tokyo
Chair, GNMM Task Force II

11:00 A.M. **Commercializing the Excess Defense Infrastructure**

What commercial opportunities exist for the peaceful application of excess nuclear weapons–related technology, materials, facilities, and specialists? What companies have already taken a leading role in this area? Is there room in the nuclear industry for further action?

Roger Howsley
Head of Security Safeguards and International Affairs, British Nuclear Fuels Limited
Chair, GNMM Task Force III

11:45 A.M. **A View from the Hill**

Pete V. Domenici (R-N.Mex.)
U.S. Senate

J. Robert Kerrey (D-Nebr.)
U.S. Senate

12:30 P.M. **Buffet Lunch**

Keynote Address—A View from the Outside

James R. Schlesinger
Former Secretary of Energy
Former Secretary of Defense
Former Director of Central Intelligence
Counselor and Trustee, CSIS

1:45 P.M. **A View from the Administration**

Ernest J. Moniz
Under Secretary of Energy

2:30 P.M. **Transparency**

What is transparency? Can it ensure that nuclear materials are safely, securely, and legitimately used throughout the world?

Roger L. Hagengruber
Senior Vice President, National Security and Arms Control Division, Sandia National Laboratories
Chair, GNMM Task Force IV

3:30 p.m. **U.S. Domestic Infrastructure and the Emerging Nuclear Era**

Do current trends and policies regarding the research, development, and use of nuclear technology enable the United States to lead?

John J. Taylor
Former Vice President of the Nuclear Power Group, Electric Power Research Institute
Chair, GNMM Task Force V

4:15 p.m. **Closing Remarks**

Robert E. Ebel
Director, Energy and National Security Program, CSIS

Looking Ahead

William C. Potter
Institute Professor and Director, Center for Non-Proliferation Studies, Monterey Institute of International Studies

4:45 p.m. **Adjourn**

About the Speakers

Matthew Bunn. Mr. Bunn is assistant director of the Science, Technology, and Public Policy Program in the Belfer Center for Science and International Affairs at Harvard University's John F. Kennedy School of Government. From 1994 to 1996, he served as an adviser to the White House Office of Science and Technology Policy, where he took part in a wide range of U.S.-Russian negotiations relating to security, monitoring, and disposition of weapons-usable nuclear materials. Previously, Mr. Bunn directed the Management and Disposition of Excess Weapons Plutonium study conducted by the National Academy of Sciences and served as editor of *Arms Control Today,* published by the Arms Control Association. He received his bachelor's and master's degrees in political science from the Massachusetts Institute of Technology.

Pete V. Domenici. With his reelection in 1996, Senator Domenici (R-N.Mex.) became the first New Mexican elected to serve five full six-year terms in the U.S. Senate. He is chairman of the Committee on the Budget and also serves on the Committee on Appropriations, Committee on Energy and Natural Resources, Governmental Affairs Committee, and the Select Committee on Indian Affairs, as well as a number of subcommittees. As chairman of the Energy and Water Appropriations Subcommittee that funds the DOE, Senator Domenici has worked to bolster the evolving mission of the national laboratories from weapons research and production to stewardship over the nation's existing nuclear weapons stockpile. He holds an education degree from the University of New Mexico and a law degree from the University of Denver.

Robert E. Ebel. Mr. Ebel is currently director of the Energy Program at CSIS. Before joining CSIS, he was vice president for international affairs at ENSERCH Corporation. Previously, he held positions in the Central Intelligence Agency, the Department of the Interior, and the Federal Energy Agency. Mr. Ebel has traveled widely in the former Soviet Union, including as a member of the first U.S. oil delegation in 1960, on a mission to inspect the new oil fields of Western Siberia in 1970, and on a team of experts examining Russia's long-term energy strategy in 1994. He is the author of *Chernobyl and its Aftermath, Energy Choices in Russia,* and *Energy in the Near Abroad.* Mr. Ebel holds a B.S. in petroleum geology from Texas Tech University and an M.A. in international relations from Syracuse University.

ROGER L. HAGENGRUBER. Dr. Hagengruber is senior vice president of the National Security and Arms Control Division and manager of the Non-Proliferation and Materials Management Strategic Business Unit for Sandia National Laboratories (SNL), where he has held a number of key positions since 1972. Before joining SNL, he taught courses in arms control and international conflict as an adjunct professor of political science at the University of New Mexico. In addition, Dr. Hagengruber has served in a variety of government assignments, including four tours in Geneva as a member of U.S. arms control negotiating teams. In 1979, he was appointed the U.S. expert to an international forum on new weapons of mass destruction. Dr. Hagengruber received his Ph.D. in experimental nuclear physics from the University of Wisconsin.

ROGER HOWSLEY. Dr. Howsley is head of Security Safeguards and International Affairs for British Nuclear Fuels Limited, which provides high-quality, cost-effective integrated nuclear products and services to customers throughout the world. The company's expertise spans fuel manufacture and uranium procurement through to recycling used fuel, transporting radioactive materials, engineering, waste management, and decommissioning. Dr. Howsley has extensive experience in working with the International Atomic Energy Agency and EURATOM and with national police forces and security organizations. His professional interests include the climate change debate and use of nuclear power in mitigating increases in atmospheric carbon dioxide. He holds a first class honors degree and doctorate in life sciences from the University of Liverpool.

J. ROBERT KERREY. Senator Kerrey (D-Nebr.) is the senior U.S. senator from Nebraska. He serves as vice chairman of the Select Committee on Intelligence and is also a member of the Finance Committee and the Agriculture Committee. Previously, Senator Kerrey served as cochair of the National Commission on Restructuring the Internal Revenue Service and as a member of the Committee on Appropriations. Before his election to the Senate, he served one term as governor of the state of Nebraska. Senator Kerrey is a former member of the elite navy SEAL team and is a highly decorated Vietnam veteran. He is currently the only member of Congress to have received the congressional Medal of Honor, America's highest military honor. Senator Kerrey is a graduate of the University of Nebraska.

ERNEST J. MONIZ. Dr. Moniz was confirmed in 1997 as under secretary of energy, where he oversees the Department of Energy's research and development portfolio, the national laboratory system, and national security programs, including stockpile stewardship and nonproliferation. Before joining the Department of Energy, Dr. Moniz was professor of physics at the Massachusetts Institute of Technology. Prior to that, he served as the associate director for science in the Office of Science and Technology Policy in the Executive Office of the President. In addition, Dr. Moniz has served numerous universities, national laboratories, professional societies, and government agencies in advisory roles. He holds a Ph.D. in theoretical physics from Stanford University.

Sam Nunn. Former Senator Nunn first entered politics as a member of the Georgia state House of Representatives in 1968. He was elected to the U.S. Senate in 1972 and served four terms. During his tenure, Senator Nunn served as chairman of the Armed Services Committee and the Permanent Subcommittee on Investigations. He coauthored legislation creating the Cooperative Threat Reduction Program, also known as the Nunn–Lugar program, which provides incentives for the former Soviet republics to dismantle and safely handle their nuclear arsenals. Senator Nunn graduated with honors from Emory Law School. He is currently a senior partner in the law firm of King and Spalding, where he is focusing his practice on international and corporate matters. In addition, Senator Nunn serves as chair of the board of trustees at CSIS.

William C. Potter. Dr. Potter is institute professor and director of the Center for Nonproliferation Studies at the Monterey Institute of International Studies (MIIS). He also directs the MIIS Center for Russian and Eurasian Studies. In addition, Dr. Potter serves on the United Nations advisory board on disarmament matters, the board of trustees of the United Nations Institute for Disarmament Research, and the international advisory board of the Center for Policy Studies in Russia. He was an adviser to the delegation of Kyrgyzstan to the 1995 Nuclear Non-Proliferation Treaty (NPT) review and extension conference and to the 1997, 1998, and 1999 sessions of the NPT preparatory committee. Dr. Potter has written three books, including *Nuclear Profiles of the Soviet Successor States.* He holds a Ph.D. in political science from the University of Michigan.

James R. Schlesinger. Dr. Schlesinger is senior adviser to the investment banking firm of Lehman Brothers and chairman of the board of trustees of the MITRE Corporation. He also serves as counselor and trustee at CSIS. In 1984, he was vice chairman of the president's Blue Ribbon Task Group on Nuclear Weapons Program Management. He served as assistant to the president to establish the Department of Energy and then became the first secretary of energy in 1977. In addition, Dr. Schlesinger served as secretary of defense, director of central intelligence, chairman of the Atomic Energy Commission, and assistant director of the Bureau of the Budget (later the Office of Management and Budget). He is the author of *The Political Economy of National Security and America at Century's End.* Dr. Schlesinger received his Ph.D. from Harvard University.

Atsuyuki Suzuki. Dr. Suzuki is a professor of nuclear engineering in the Department of Quantum Engineering and Systems Science at the University of Tokyo, where he has been a member of the faculty since 1971. Dr. Suzuki has served on advisory committees and boards for a number of agencies, including the Japanese Atomic Energy Commission, Nuclear Safety Commission, Nuclear Safety Research Association, Institute of Nuclear Fuel Cycle Policy, and Nuclear Safety Technology Center. He is the author or coauthor of more than 200 papers and 30 books and is the editor of *Nuclear Technology* for the American Nuclear Society and the international journal, *Radioactive Waste Management and Environmental Restoration.* He received his Ph.D. in nuclear engineering from the University of Tokyo.

JOHN J. TAYLOR. Mr. Taylor is a consultant on nuclear power. Until 1995, he was vice president of the Nuclear Power Group at the Electric Power Research Institute, where he was responsible for nuclear power research and development sponsored by U.S. and international utilities. Previously, he was vice president at Westinghouse Electric Corporation, where he participated in development of the first nuclear-powered submarines and ships for the navy and nuclear steam supply systems for generating electricity. He is a member of the Nuclear Energy Research Advisory Committee at the Department of Energy, the Radioactive Waste Management Board of the National Research Council, and the U.S.-Russian Independent Scientific Commission on the Disposition of Excess Weapons Plutonium. He received his master's degree from the University of Notre Dame.